I0064516

100 INTEGRALS

LICENSE, DISCLAIMER OF LIABILITY, AND LIMITED WARRANTY

By purchasing or using this book (the "Work"), you agree that this license grants permission to use the contents contained herein, but does not give you the right of ownership to any of the textual content in the book or ownership to any of the information or products contained in it. *This license does not permit uploading of the Work onto the Internet or on a network (of any kind) without the written consent of the Publisher.* Duplication or dissemination of any text, code, simulations, images, etc. contained herein is limited to and subject to licensing terms for the respective products, and permission must be obtained from the Publisher or the owner of the content, etc., in order to reproduce or network any portion of the textual material (in any media) that is contained in the Work.

MERCURY LEARNING AND INFORMATION ("MLI" or "the Publisher") and anyone involved in the creation, writing, or production of the companion disc, accompanying algorithms, code, or computer programs ("the software"), and any accompanying Web site or software of the Work, cannot and do not warrant the performance or results that might be obtained by using the contents of the Work. The author, developers, and the Publisher have used their best efforts to insure the accuracy and functionality of the textual material and/or programs contained in this package; we, however, make no warranty of any kind, express or implied, regarding the performance of these contents or programs. The Work is sold "as is" without warranty (except for defective materials used in manufacturing the book or due to faulty workmanship).

The author, developers, and the publisher of any accompanying content, and anyone involved in the composition, production, and manufacturing of this work will not be liable for damages of any kind arising out of the use of (or the inability to use) the algorithms, source code, computer programs, or textual material contained in this publication. This includes, but is not limited to, loss of revenue or profit, or other incidental, physical, or consequential damages arising out of the use of this Work.

The sole remedy in the event of a claim of any kind is expressly limited to replacement of the book, and only at the discretion of the Publisher. The use of "implied warranty" and certain "exclusions" vary from state to state, and might not apply to the purchaser of this product.

100 INTEGRALS

Solutions and Engineering Applications

MEHRZAD TABATABAIAN, PhD, PEng

MERCURY LEARNING AND INFORMATION
Boston, Massachusetts

Copyright ©2023 by MERCURY LEARNING AND INFORMATION LLC. All rights reserved.
An imprint of De Gruyter Inc.

This publication, portions of it, or any accompanying software may not be reproduced in any way, stored in a retrieval system of any type, or transmitted by any means, media, electronic display or mechanical display, including, but not limited to, photocopy, recording, Internet postings, or scanning, without prior permission in writing from the publisher.

Publisher: David Pallai
MERCURY LEARNING AND INFORMATION
121 High Street, 3rd Floor
Boston, MA 02110
info@merclearning.com
www.merclearning.com
(800) 232-0223

M. Tabatabaian. *100 Integrals: Solutions and Engineering Applications.*
ISBN: 978-1-68392-967-3

The publisher recognizes and respects all marks used by companies, manufacturers, and developers as a means to distinguish their products. All brand names and product names mentioned in this book are trademarks or service marks of their respective companies. Any omission or misuse (of any kind) of service marks or trademarks, etc. is not an attempt to infringe on the property of others.

Library of Congress Control Number: 2023942788

232425321 This book is printed on acid-free paper in the United States of America.

Our titles are available for adoption, license, or bulk purchase by institutions, corporations, etc. For additional information, please contact the Customer Service Dept. at 800-232-0223 (toll free).

All of our titles are available in digital format at *www.academiccourseware.com* and other digital vendors. The sole obligation of MERCURY LEARNING AND INFORMATION to the purchaser is to replace the book, based on defective materials or faulty workmanship, but not based on the operation or functionality of the product.

To the relentless explorers of integral calculus, may this textbook become your trusted companion, guiding you through the labyrinth of calculations, modelling, and simulations in engineering and technical fields.

CONTENTS

PREFACE

This monograph contains a collection of integrals, some more challenging than others, with their worked-out solutions as indefinite integrals. The integrals were randomly selected, modified, or designed with the condition of having closed forms solutions with common functions. This list is meant for helping readers in practicing and getting hints for working out solutions to similar integrals that they might encounter. Readers might want to add their own favorite integrals to this list. By no means is this a comprehensive list of integrals, as many authors have created such lists ([1], [2], [3], [4], [5], [6], [7], [8], [9], [10]).

The *exercise* of integration operation is a mind stimulating activity, as it requires the knowledge of certain mathematical techniques and the discovery of tricks and short cuts while solving them. The latter feature makes integration different and more enjoyable compared to other topics in calculus (e.g., differentiation and algebraic manipulation).

In addition, we present the application of some integrals in engineering related topics. For example, nonuniform loading, hydrostatic force, moment of inertia, polar moment of inertia, etc.

We introduce an up-to-date online software tool, WolframAlpha[1] that can be used for comparing our answers for the integrals listed. However, readers may want to update to WolframAlpha Pro for recovering some of the integrals' step-by-step solutions. Interested readers might like to try this tool for their selected integrals from the list. However, please note that sometimes equivalent results are provided by this tool for the same integral as input when compared to the results presented in this volume. In addition, similar CAS (computer algebra system) tools like Maple[2], Mathematica[3], or Mathcad[4] may also be employed.

Finally, readers should be aware that some integrals may have alternative equivalent solutions rather than a corresponding unique one. Thus worked-out solutions may be different, but equivalent, to those provided by some online CAS tools.

Mehrzad Tabatabaian, PhD, PEng
Vancouver, B.C.
July 22, 2023

---

1   *https://www.wolframalpha.com/calculators/integral-calculator/*

2   *https://www.maplesoft.com*

3   *https://www.wolfram.com/mathematica/*

4   *https://www.mathcad.com*

ABOUT THE AUTHOR

Dr. Mehrzad Tabatabaian is a faculty member in the Mechanical Engineering Department, School of Energy at the British Columbia Institute of Technology. He has several years of teaching and industry experience. Dr. Tabatabaian is currently Chair of the BCIT School of Energy Research Committee. He has published several papers in scientific journals and conferences, and he has written textbooks on multiphysics and turbulent flow modelling, advanced thermodynamics, tensor analysis, direct energy conversion, and Bond Graph modelling method. He holds several registered patents in the energy field resulting from years of research activities.

Dr. Tabatabaian volunteered to help establish the Energy Efficiency and Renewable Energy Division (EERED), a new division at Engineers and Geoscientists British Columbia (EGBC).

Mehrzad Tabatabaian received his BEng from Sharif University of Technology (former AUT) and advanced degrees from McGill University (MEng and PhD). He has been an active academic, professor, and engineer in leading alternative energy, oil, and gas industries. Mehrzad has also a Leadership Certificate from the University of Alberta and holds an EGBC P.Eng. License.

LIST OF SELECTED INTEGRALS WITH THEIR STEP-BY-STEP SOLUTIONS

TABLE 1 List of selected integrals for Part 1

Integral		Integral	
1	$\int x\ln^3 x\,dx$	19	$\int \ln\left(1+\sqrt{1+\frac{1}{x^2}}\right)dx$
2	$\int e^x \sqrt{1+e^{2x}}\,dx$	20	$\int \operatorname{csch}^{-1}x\,dx$
3	$\int \frac{\cos^3 x}{\sqrt{\sin x}}\,dx$	21	$\int \frac{1}{\sin x + \cos x}\,dx$
4	$\int \ln\left(\sqrt{x}+\sqrt{1+x}\right)dx$	22	$\int 2^{\sqrt{-x}}\,dx$
5	$\int \frac{1}{\cos x}\,dx$	23	$\int \ln\left(1+\sqrt{1+x^2}\right)dx$
6	$\int \frac{x}{\left(\sin x + \cos x\right)^2}\,dx$	24	$\int \sin^6 x \cos^5 x\,dx$
7	$\int \frac{x}{\left(7x+1\right)^{17}}\,dx$	25	$\int x^m \ln^n x\,dx$
8	$\int \frac{x^3}{\left(1+x^2\right)^2}\,dx$	26	$\int e^x \sin^{-1}\left(e^x\right)dx$
9	$\int \frac{x}{8+4x^2+x^4}\,dx$	27	$\int x\sqrt{\frac{1-x^2}{1+x^2}}\,dx$
10	$\int \frac{4x^2+x+1}{4x^3+x}\,dx$	28	$\int x\sqrt{x}\tan^{-1}\sqrt{x}\,dx$
11	$\int \frac{x^4+1}{x^2+2}\,dx$	29	$\int \frac{\tan^{-1}x}{\left(x-1\right)^3}\,dx$
12	$\int \frac{1}{5+4\cos x}\,dx$	30	$\int \frac{\sin^{-1}x}{x^2}\,dx$
13	$\int \sec^4 x\,dx$	31	$\int x\sec^{-1}x\,dx$
14	$\int \frac{1}{\sqrt{\left(1+x^2\right)^5}}\,dx$	32	$\int \sec^{-1}\sqrt{x}\,dx$
15	$\int \sqrt{\tan x}\,dx$	33	$\int x^2 \tan^{-1}x\,dx$
16	$\int xe^x \sin x\,dx$	34	$\int \frac{2x+3}{\sqrt{3+6x-9x^2}}\,dx$
17	$\int \frac{\ln^3 x}{x^3}\,dx$	35	$\int \frac{3x+2}{\left(x^2+4\right)\sqrt{x^2+4}}\,dx$
18	$\int \ln\left(1+\sqrt{x}\right)dx$	36	$\int \frac{2x^2-5x-1}{x^3-2x^2-x+2}\,dx$

Integral		Integral	
37	$\int \dfrac{2x^3 + 3x^2 + 4}{(x+1)^4}\,dx$	54	$\int \dfrac{1}{\left(x+\dfrac{1}{x}\right)^2}\,dx$
38	$\int \dfrac{x-1}{(x+1)\sqrt{x^3+x^2+x}}\,dx$	55	$\int e^{2x}\tan^{-1}(e^x)\,dx$
39	$\int \dfrac{x^4+4x^3+6x^2+4x+1}{x^3-3x^2+3x-1}\,dx$	56	$\int \dfrac{1}{x\sqrt{6x-x^2}}\,dx$
40	$\int \dfrac{1}{\sqrt{x}\left(\sqrt[4]{x}+1\right)^{10}}\,dx$	57	$\int \tan^3 x\,dx$
41	$\int (1+\ln x)\ln(\ln x)\,dx$	58	$\int \dfrac{1}{4+5\cos x}\,dx$
42	$\int \ln\left(x^2+x+1\right)\,dx$	59	$\int \dfrac{\tan^3(1+\ln x)}{x}\,dx$
43	$\int \cos x\sqrt{4-\sin^2 x}\,dx$	60	$\int \left(\dfrac{\sin 2x\sin 3x}{\sin x\sin 6x}\right)^2\,dx$
44	$\int \dfrac{\cos x}{\sin^2 x-3\sin x+2}\,dx$	61	$\int \dfrac{\sin^3 x}{\sqrt{\cos x}}\,dx$
45	$\int \dfrac{1}{\left(1+\sqrt{x}\right)\sqrt{x-x^2}}\,dx$	62	$\int \dfrac{5x+31}{3x^2-4x+11}\,dx$
46	$\int \dfrac{1+2x^2}{x^5\left(1+x^2\right)^3}\,dx$	63	$\int \dfrac{3x^5-x^4+2x^3-12x^2-2x+1}{\left(x^3-1\right)^2}\,dx$
47	$\int \sin^6 x\cos^5 x\,dx$, alternative solution	64	$\int \dfrac{4x^3-x+1}{x^3+1}\,dx$
48	$\int x^2 e^{x^{3/2}}\,dx$	65	$\int \sin^{-1} x\ln x\,dx$
49	$\int \dfrac{\tan^3 x}{\cos^3 x}\,dx$	66	$\int \ln(\sin x)\sqrt{1+\sin x}\,dx$
50	$\int \dfrac{\sqrt[3]{\tan x}}{(\sin x+\cos x)^2}\,dx$	67	$\int \dfrac{x^3 e^{\sin^{-1} x}}{\sqrt{1-x^2}}\,dx$
51	$\int \left(\dfrac{1}{\ln x}-\dfrac{1}{\ln^2 x}\right)\,dx$	68	$\int \dfrac{\tan x}{\sqrt{1+\sec^3 x}}\,dx$
52	$\int \left[\sin(x+\sin x)-\sin(x-\sin x)\right]\,dx$	69	$\int \dfrac{x\ln\left(x+\sqrt{x^2-1}\right)}{\sqrt{x^2-1}}\,dx$
53	$\int \dfrac{x}{1-5x^2}\sqrt{\dfrac{2}{1+5x^2}-1}\,dx$	70	$\int \sin^{-1}\left(\sqrt{1+x}\right)\,dx$

Integral		Integral	
71	$\int \dfrac{1}{2+2\sin x + \cos x}\,dx$	86	$\int \dfrac{1}{x^4+4}\,dx$
72	$\int \dfrac{\sec^2 x}{\tan^2 x + 2\tan x + 2}\,dx$	87	$\int \dfrac{x^5}{\left(4x^2+4\right)^{5/2}}\,dx$
73	$\int \sqrt{x^2+x+1}\,dx$	88	$\int \dfrac{1}{x^4+1}\,dx$
74	$\int \dfrac{x}{\left(x^2+2x+2\right)^2}\,dx$	89	$\int \sin^2\left(\ln x\right)dx$
75	$\int \dfrac{1}{x^3\sqrt{x^2-9}}\,dx$	90	$\int \sin x \sin 2x \sin 3x\,dx$
76	$\int \sin x \tan^{-1}\sqrt{\sec x - 1}\,dx$	91	$\int \sqrt{1+x^2}\,dx$
77	$\int \dfrac{1}{\sqrt{x\sqrt{x}-x^2}}\,dx$	92	$\int \dfrac{\ln x \cos x - \dfrac{1}{x}\sin x}{\left(\ln x\right)^2}\,dx$
78	$\int x^3\left(\ln\left(x^{\ln x - 6}\right)+5\right)dx$	93	$\int \dfrac{2e^{2x}-e^x}{\sqrt{3e^{2x}-6e^x-1}}\,dx$
79	$\int \tan^{-1}\left(\sqrt{x+1}-\sqrt{x}\right)dx$	94	$\int \left(\dfrac{x^4}{1+x^6}\right)^2 dx$
80	$\int \dfrac{x^9}{x^{20}-48x^{10}+575}\,dx$	95	$\int \dfrac{x^3 e^{x^2}}{\left(1+x^2\right)^2}\,dx$
81	$\int \sin^{-1}x\,dx$	96	$\int \dfrac{1}{1+\sqrt{\sqrt{x}}}\,dx$
82	$\int \tan^{-1}x\,dx$	97	$\int \left(\sin^{-1}x\right)^2 dx$
83	$\int \sinh^{-1}x\,dx$	98	$\int e^{x^x}\left(1+\ln x\right)x^{2x}\,dx$
84	$\int \tanh^{-1}x\,dx$	99	$\int x\pi^{\ln x}\,dx$
85	$\int \cos^{-1}\left(\dfrac{1}{x}\right)dx$	100	$\int \dfrac{\ln\left(\sin x\right)}{1+\sin x}\,dx$

INTEGRAL 1

Problem

$$\int x ln^3 x \, dx$$

Solution:

$$\frac{x^2}{8}\left(4ln^3 x - 6ln^2 x + 6lnx - 3\right) + constant$$

Techniques used:

Change of variables, Integration by parts

Step-by-step solution:

Let $lnx = z$, then we have $dx = xdz$, $x = e^z$. After rewriting the integral in terms of variable z, we get $\int x ln^3 x \, dx = \int e^{2z} z^3 dz$. Integrating by parts gives, $\int \underset{dg}{e^{2z}} \underset{f}{z^3} \, dz = \frac{1}{2} e^{2z} z^3 - \frac{1}{2}\int e^{2z}\left(3z^2\right)dz$. Performing the integration by parts technique, two times, on the second term gives $\int e^{2z} z^3 dz = \frac{1}{2} e^{2z} z^3 - \frac{3}{4} e^{2z} z^2 + \frac{3}{2}\int e^{2z} z \, dz = \frac{1}{2} e^{2z} z^3 - \frac{3}{4} e^{2z} z^2 + \frac{3}{4} e^{2z} z - \frac{3}{8} e^{2z}$. Factoring out $e^{2z} = x^2$ and substituting for $z = lnx$, we can rewrite the result in terms of the original variable x as $\frac{1}{8} e^{2z}\left(4z^3 - 6z^2 + 6z - 3\right) = \frac{x^2}{8}\left(4ln^3 x - 6ln^2 x + 6lnx - 3\right)$.

INTEGRAL 2

Problem

$$\int e^x \sqrt{1 + e^{2x}}\, dx$$

Solution:

$$\frac{1}{2}\left[\sinh^{-1}\left(e^x\right) + e^x \sqrt{1 + e^{2x}} \right] + constant$$

Techniques used:

Change of variables, Trigonometric identities

Step-by-step solution:

Let $e^x = z \Rightarrow dx = \dfrac{dz}{z}$. Rewriting the integral in terms of variable z gives $\int e^x \sqrt{1 + e^{2x}}\, dx = \int z\sqrt{1 + z^2}\,\dfrac{dz}{z} = \int \sqrt{1 + z^2}\, dz$. Now, let $z = \sinh u \Rightarrow dz = \cosh u\, du$. Rewriting the new integral in terms of variable u, gives $\int \sqrt{1 + z^2}\, dz = \int \sqrt{1 + \sinh^2 u}\,\cosh u\, du$. But, using the $\cosh^2 u - \sinh^2 u = 1$ identity we get $\int \sqrt{1 + \sinh^2 u}\,\cosh u\, du = \int \sqrt{\cosh^2 u}\,\cosh u\, du = \int \cosh^2 u\, du$. Now, using the $\cosh^2 u = (1 + \cosh 2u)/2$ identity, we get $\int \cosh^2 u\, du = \int \dfrac{1 + \cosh 2u}{2}\, du$. Expanding the integrand and integrate each term gives $\int \dfrac{1 + \cosh 2u}{2}\, du = \int \dfrac{1}{2}\, du + \dfrac{1}{2}\int \cosh 2u\, du = \dfrac{u}{2} + \dfrac{\sinh 2u}{4} = \dfrac{u}{2} + \dfrac{\sinh 2u}{4} = \dfrac{u}{2} +$. Rewriting this expression in terms of variable z and subsequently x, we get $\dfrac{u}{2} + \dfrac{2\sinh u \cosh u}{2} = \dfrac{\sinh^{-1} z}{2} + \dfrac{z\sqrt{1 + z^2}}{2}$

$$= \frac{1}{2}\left(\sinh^{-1}\left(e^x\right) + e^x \sqrt{1 + e^{2x}} \right).$$

INTEGRAL 3

Problem

$$\int \frac{\cos^3 x}{\sqrt{\sin x}}\, dx$$

Solution:

$$\frac{2}{5}\sqrt{\sin x}\left(1 + \cos^2 x\right) + \text{constant}$$

Techniques used:

Change of variables, Trigonometric identities

Step-by-step solution:

Let $\sin x = z \Rightarrow \cos x\, dx = dz$. Rewriting the integral in terms of variable z gives $\int \dfrac{\cos^2 x}{\sqrt{\sin x}}\cos x\, dx = \int \dfrac{1-z^2}{\sqrt{z}}dz$. Now, let $\sqrt{z} = u \Rightarrow dz = 2u\, du$.

Rewriting the new integral in terms of variable u gives $\int \dfrac{1-z^2}{\sqrt{z}}dz = 2\int \dfrac{1-u^4}{u}u\, du = 2u - \dfrac{2}{5}u^5$. Substituting back and rewriting the answer in terms of variable x, after some manipulations, gives

$$2u - \frac{2}{5}u^5 = 2\sqrt{\sin x} - \frac{2}{5}\left(\sqrt{\sin x}\right)^5 = 2\sqrt{\sin x}\left(1 - \frac{1}{5}\sin^2 x\right) = 2\sqrt{\sin x}$$

$$\left(1 - \frac{1}{5}\left(1 - \cos^2 x\right)\right) = \frac{2}{5}\sqrt{\sin x}\left(4 + \cos^2 x\right).$$

INTEGRAL 4

Problem

$$\int \ln\left(\sqrt{x} + \sqrt{1+x}\right)dx$$

Solution:

$$\frac{1}{2}\left[\sinh^{-1}\sqrt{x} - \sqrt{x(1+x)} + 2x\ln\left(\sqrt{x} + \sqrt{1+x}\right)\right] + constant$$

Techniques used:

Change of variables, Integration by parts, Trigonometric identities

Step-by-step solution:

We use integration by parts technique with considering $\int dx = x$. Therefore, we can write $\int \ln\left(\sqrt{x} + \sqrt{1+x}\right)dx = \ln\left(\sqrt{x} + \sqrt{1+x}\right)\int dx -$

$\int x \frac{d}{dx}\ln\left(\sqrt{x} + \sqrt{1+x}\right)dx$. Performing the integration gives $x\ln\left(\sqrt{x} +\right.$

$$\left.\sqrt{1+x}\right) - \int x \frac{\dfrac{1}{2\sqrt{x}} + \dfrac{1}{2\sqrt{1+x}}}{\sqrt{x}+\sqrt{1+x}} = x\ln\left(\sqrt{x} + \sqrt{1+x}\right) - \frac{1}{2}\int \frac{\sqrt{x}}{\sqrt{1+x}}dx.$$

Now, we let $\sqrt{x} = z \Rightarrow dx = 2zdz$. Rewriting the new integral in terms of the variable z, gives $-\frac{1}{2}\int \frac{\sqrt{x}}{\sqrt{1+x}}dx = -\int \frac{z^2}{\sqrt{1+z^2}}dz$. But

using $\frac{d}{dx}\left(\sqrt{1+z^2}\right) = \frac{z}{\sqrt{1+z^2}}$, and the integration by parts technique

we can write $-\int \frac{z^2}{\sqrt{1+z^2}} = -\int \frac{z}{\sqrt{1+z^2}}zdz = -z\sqrt{1+z^2} + \int \sqrt{1+z^2}\,dz.$

Now, we let $z = \sinh u \Rightarrow dz = \cosh u du$. Hence, the latter integral can be written as $\int \sqrt{1+z^2}\,dz = \int \sqrt{1+\sinh^2 u}\,\cosh u\,du = \int \cosh^2 u\,du$, after using the $\cosh^2 u - \sinh^2 u = 1$ identity. But $\cosh^2 u = \frac{1}{2}(1+\cosh 2u)$ and after rewriting the last integral we have

$$\int \cosh^2 u\,du = \frac{1}{2}\int(1+\cosh 2u)\,du = \frac{1}{2}u + \frac{1}{4}\sinh 2u = \frac{1}{2}u + \frac{1}{2}\sinh u$$

$\cosh u$. After collecting all related answers and rewrite them in terms of the original variable x, we get the solution as shown.

INTEGRAL 5

Problem

$$\int \frac{1}{\cos x}\, dx$$

Solution:

$$\ln\left[\frac{1 + \tan\left(\dfrac{x}{2}\right)}{1 - \tan\left(\dfrac{x}{2}\right)}\right] + constant$$

Techniques used:

Integration by parts, Partial fractions, Trigonometric identities

Step-by-step solution:

We rewrite the integral using the trigonometric identity $\cos x = \cos^2 \dfrac{x}{2} - \sin^2 \dfrac{x}{2}$. Therefore, $\displaystyle\int \frac{1}{\cos x}\, dx = \int \frac{1}{\cos^2 \dfrac{x}{2} - \sin^2 \dfrac{x}{2}}$

$dx = \displaystyle\int \frac{1}{\left(\cos\dfrac{x}{2} - \sin\dfrac{x}{2}\right)\left(\cos\dfrac{x}{2} + \sin\dfrac{x}{2}\right)}\, dx$. But the integrand can

be written as, using the partial fractions technique,

$$\frac{1}{\left(\cos\dfrac{x}{2} - \sin\dfrac{x}{2}\right)\left(\cos\dfrac{x}{2} + \sin\dfrac{x}{2}\right)} = \frac{a}{\cos\dfrac{x}{2} - \sin\dfrac{x}{2}} + \frac{b}{\cos\dfrac{x}{2} + \sin\dfrac{x}{2}} =$$

$$\frac{(a-b)\sin\dfrac{x}{2} + (a+b)\cos\dfrac{x}{2}}{\left(\cos\dfrac{x}{2} - \sin\dfrac{x}{2}\right)\left(\cos\dfrac{x}{2} + \sin\dfrac{x}{2}\right)}.$$ Therefore, to have equality valid,

the constants a and b should satisfy $(a-b)\sin$

$\dfrac{x}{2} + (a+b)\cos\dfrac{x}{2} = 1$. Or, $a - b = \sin\dfrac{x}{2}$ and $(a+b) = \cos\dfrac{x}{2}$.

Solving for a and b gives, $a = \dfrac{1}{2}\left(\cos\dfrac{x}{2} + \sin\dfrac{x}{2}\right)$ and $b = \dfrac{1}{2}\left(\cos\dfrac{x}{2} - \sin\dfrac{x}{2}\right)$. Therefore, the original integral can be

written as $\displaystyle\int \dfrac{1}{\left(\cos\dfrac{x}{2} - \sin\dfrac{x}{2}\right)\left(\cos\dfrac{x}{2} + \sin\dfrac{x}{2}\right)}\,dx = \dfrac{1}{2}\int \dfrac{\cos\dfrac{x}{2} + \sin\dfrac{x}{2}}{\cos\dfrac{x}{2} - \sin\dfrac{x}{2}}\,dx + $

$\dfrac{1}{2}\displaystyle\int \dfrac{\cos\dfrac{x}{2} - \sin\dfrac{x}{2}}{\cos\dfrac{x}{2} + \sin\dfrac{x}{2}}\,dx.$ But $\dfrac{d}{dx}\left(\cos\dfrac{x}{2} - \sin\dfrac{x}{2}\right) = -\left(\cos\dfrac{x}{2} + \sin\dfrac{x}{2}\right)$ and

$\dfrac{d}{dx}\left(\cos\dfrac{x}{2} + \sin\dfrac{x}{2}\right) = \left(\cos\dfrac{x}{2} - \sin\dfrac{x}{2}\right).$ Using these relations and recall-

ing that $\dfrac{d}{dz}\ln z = \dfrac{dz}{z}$, we can arrive at the final answer for the integral

as $\dfrac{1}{2}\displaystyle\int \dfrac{\cos\dfrac{x}{2} + \sin\dfrac{x}{2}}{\cos\dfrac{x}{2} - \sin\dfrac{x}{2}}\,dx + \dfrac{1}{2}\int \dfrac{\cos\dfrac{x}{2} - \sin\dfrac{x}{2}}{\cos\dfrac{x}{2} + \sin\dfrac{x}{2}}\,dx = -\ln\left(\cos\dfrac{x}{2} - \sin\dfrac{x}{2}\right) + $

$\ln\left(\cos\dfrac{x}{2} + \sin\dfrac{x}{2}\right) = \ln\left(\dfrac{\cos\dfrac{x}{2} + \sin\dfrac{x}{2}}{\cos\dfrac{x}{2} - \sin\dfrac{x}{2}}\right).$ To simplify, after dividing the

numerator and denominator of the argument of the logarithm by

$\cos\dfrac{x}{2}$, we get $\displaystyle\int \dfrac{1}{\cos x}\,dx = \ln\left(\dfrac{1 + \tan\dfrac{x}{2}}{1 - \tan\dfrac{x}{2}}\right).$

Short-cut solution:

An alternative solution can be obtained as follow:

Let $\sin x = z \Rightarrow dx = \dfrac{dz}{\cos x}$. Rewriting the integral gives

$\displaystyle\int \dfrac{1}{\cos x}\,dx = \int \dfrac{1}{\cos^2 x}\,dz = \int \dfrac{1}{1 - z^2}\,dz.$ But $\displaystyle\int \dfrac{1}{1 - z^2}\,dz = \tanh^{-1} z$, where

in terms of the original variable x the answer reads $\tanh^{-1}(\sin x)$.

INTEGRAL 6

Problem

$$\int \frac{x}{\left(\sin x + \cos x\right)^2} \, dx$$

Solution:

$$\frac{1}{2}\left[x + \ln\left(\sin x + \cos x\right) - \frac{2x\cos x}{\sin x + \cos x} \right] + constant$$

Techniques used:

Integration by parts, Partial fractions, Trigonometric identities

Step-by-step solution:

We apply the integration by parts technique to the integral, Or $\int \frac{x}{\left(\sin x + \cos x\right)^2} \, dx = x\int \frac{dx}{\left(\sin x + \cos x\right)^2} - \int x \int \frac{dx}{\left(\sin x + \cos x\right)^2}$. This requires calculating the $\int \frac{dx}{\left(\sin x + \cos x\right)^2}$. But, by inspection the derivative of an expression, like $\left(\frac{f(x)}{\sin x + \cos x} \right)$ would contain the correct expression as exists in the denominator of the integrands (i.e., $\left(\sin x + \cos x\right)^2$). Assuming $f(x)$ as a polynomial function, we then have $\frac{d}{dx}\left(\frac{f(x)}{\sin x + \cos x} \right) = \frac{f'(\sin x + \cos x) - f(-\sin x + \cos x)}{\left(\sin x + \cos x\right)^2}$, with prime symbol indicating differentiation operation. To have the numerator equal to 1, the function $f(x)$ should be a sinusoidal one. Here, we let $f(x) = \cos x$. Note that selecting $f(x) = \sin x$ works as well. Therefore, $\frac{d}{dx}\left(\frac{f(x)}{\sin x + \cos x} \right) = \frac{d}{dx}\left(\frac{\cos x}{\sin x + \cos x} \right) =$

$$\frac{-\sin x(\sin x + \cos x) - \cos x(-\sin x + \cos x)}{\left(\sin x + \cos x\right)^2} = \frac{-1}{\left(\sin x + \cos x\right)^2}.$$

Therefore, we can write $\int \dfrac{1}{\left(\sin x+\cos x\right)^2}dx=-\dfrac{\cos x}{\sin x+\cos x}$.

The original integral can be written as

$$\int \dfrac{x}{\left(\sin x+\cos x\right)^2}dx=x\int \dfrac{dx}{\left(\sin x+\cos x\right)^2}-\int x\int \dfrac{dx}{\left(\sin x+\cos x\right)^2}=$$

$\dfrac{-x\cos x}{\sin x+\cos x}+\int \dfrac{\cos x}{\sin x+\cos x}dx$. But $\int \dfrac{\cos x}{\sin x+\cos x}$ can be worked

out by first dividing the integrand by $\cos x$, or

$\int \dfrac{\cos x/\cos x}{\sin x/\cos x+\cos x/\cos x}dx=\int \dfrac{1}{\tan x+1}dx$. Now, we can

manipulate the integrand in this integral as $\int \dfrac{1}{\tan x+1}dx=$

$\int \dfrac{1+\overbrace{\tan x-\tan x}^{=0}}{\tan x+1}dx=\int dx-\int \dfrac{\tan x}{\tan x+1}dx=x-\int \dfrac{\tan x+\overbrace{1-1}^{=0}}{\tan x+1}dx=$

$x-\underbrace{\int \dfrac{1}{\tan x+1}dx}+\int \dfrac{1-\tan x}{\tan x+1}dx$. Now we take the $\underbrace{-\int \dfrac{1}{\tan x+1}dx}$

from RHS to the LHS, to get $2\underbrace{\int \dfrac{1}{\tan x+1}dx}=x+\int \dfrac{1-\tan x}{\tan x+1}dx$.

But the last integrand can be written as $\int \dfrac{1-\tan x}{\tan x+1}dx=$

$\int \dfrac{\cos x-\sin x}{\sin x+\cos x}=\ln\left(\sin x+\cos x\right)$. Finally, the original integral can be

written as $\int \dfrac{x}{\left(\sin x+\cos x\right)^2}dx=\dfrac{-x\cos x}{\sin x+\cos x}+\dfrac{x}{2}+\dfrac{1}{2}\ln\left(\sin x+\cos x\right)$.

Factoring ½ and rearranging the terms gives the expression as shown above for the answer.

INTEGRAL 7

Problem

$$\int \frac{x}{(7x+1)^{17}}\, dx$$

Solution:

$$-\frac{1+112x}{11760(7x+1)^{16}} + constant$$

Techniques used:

Change of variables, Partial fractions

Step-by-step solution:

Let $7x+1 = z \Rightarrow 7dx = dz, x = \dfrac{z-1}{7}$. Writing the integral in terms of variable z, gives $\int \dfrac{x}{(7x+1)^{17}}\, dx = \int \dfrac{(z-1)/7}{z^{17}}\dfrac{dz}{7} = \dfrac{1}{49}\int\dfrac{z-1}{z^{17}}\, dz =$

$\dfrac{1}{49}\left(\int z^{-16}dz - \int z^{-17}dz\right)$. The integrals in the last expression, can

be worked to have $\int z^{-16}dz = -\dfrac{1}{15}z^{-15}$ and $\int z^{-17}dz = -\dfrac{1}{16}z^{-16}$. After

substituting and rewriting the integral in terms of original x, we

get $\dfrac{1}{49}\left(\int z^{-16}dz - \int z^{-17}dz\right) = \dfrac{1}{49}\left(-\dfrac{1}{15}z^{-15} + \dfrac{1}{16}z^{-16}\right) = \dfrac{1}{49(7x+1)^{16}}$

$\left(\dfrac{1}{16} - \dfrac{7x+1}{15}\right)$. Further simplification gives $\dfrac{1}{49(7x+1)^{16}}\left(\dfrac{1}{16} - \right.$

$\left.\dfrac{7x+1}{15}\right) = -\dfrac{1+112x}{11760(7x+1)^{16}}$, as shown in the answer.

INTEGRAL 8

Problem

$$\int \frac{x^3}{\left(1+x^2\right)^2}\,dx$$

Solution:

$$\frac{1}{2}\left[\ln\left(1+x^2\right)+\frac{1}{1+x^2}\right]+constant$$

Techniques used:

Change of variables, Partial fractions

Step-by-step solution:

By inspection and considering the denominator of the integrand, we have $\dfrac{d}{dx}\left(1+x^2\right)^2 = 4x^3 + 4x$. Therefore, we rewrite the integral as

$$\int\frac{x^3}{\left(1+x^2\right)^2}\,dx = \frac{1}{4}\int\frac{\overbrace{4x^3+4x-4x}^{=0}}{\left(1+x^2\right)^2}\,dx = \frac{1}{4}\int\frac{4x^3+4x}{\left(1+x^2\right)^2}\,dx - \int\frac{x}{\left(1+x^2\right)^2}\,dx =$$

$$\frac{1}{4}\ln\left(1+x^2\right)^2 - \int\frac{x}{\left(1+x^2\right)^2}\,dx.\ \text{To calculate the new integral, we let}$$

$1+x^2 = z \Rightarrow dx = \dfrac{dz}{2x}$. Therefore, rewriting this integral in terms of the

variable z we get $-\int\dfrac{x}{\left(1+x^2\right)^2}\,dx = -\dfrac{1}{2}\int\dfrac{x}{z^2}\dfrac{dz}{x} = -\dfrac{1}{2}\int\dfrac{dz}{z^2} = \dfrac{1}{2z}$. Back

substituting, gives the final answer in terms of the variable x as shown

$$\int\frac{x^3}{\left(1+x^2\right)^2}\,dx = \frac{1}{4}\ln\left(1+x^2\right)^2 - \int\frac{x}{\left(1+x^2\right)^2}\,dx = \frac{1}{2}\left[\ln\left(1+x^2\right)+\frac{1}{1+x^2}\right].$$

INTEGRAL 9

Problem

$$\int \frac{x}{8 + 4x^2 + x^4} \, dx$$

Solution:

$$\frac{1}{4} \tan^{-1}\left(\frac{x^2 + 2}{2}\right) + constant$$

Techniques used:

Change of variables, Trigonometric identities

Step-by-step solution:

We rewrite the denominator of the integrand as $8 + 4x^2 + x^4 = \left(x^2 + 2\right)^2 + 4$. Therefore, the integral reads as $\int \frac{x}{8 + 4x^2 + x^4} \, dx = \int \frac{x}{\left(x^2 + 2\right)^2 + 4} \, dx$. Now, let $x^2 + 2 = z \Rightarrow dx = \frac{dz}{2x}$ and rewrite the integral in terms of the variable z, or $\int \frac{x}{\left(x^2 + 2\right)^2 + 4} \, dx = \frac{1}{2} \int \frac{dz}{4 + z^2} = \frac{1}{8} \int \frac{dz}{1 + \left(z/2\right)^2}$. Now, we let $\frac{z}{2} = u \Rightarrow dz = 2du$ and write the new integral in terms of the variable u as $\frac{1}{8} \int \frac{dz}{1 + \left(z/2\right)^2} = \frac{1}{4} \int \frac{du}{1 + u^2}$. The last integral is readily equal to $\tan^{-1} u$, after back substitutions and rewriting the integral in terms of the original variable x, we get $\int \frac{x}{8 + 4x^2 + x^4} \, dx = \frac{1}{4} \tan^{-1}\left(\frac{x^2 + 2}{2}\right)$ as shown in the answer.

INTEGRAL 10

Problem

$$\int \frac{4x^2 + x + 1}{4x^3 + x} \, dx$$

Solution:

$$\ln x + \frac{1}{2} \tan^{-1}(2x) + constant$$

Techniques used:

Change of variables, Partial fractions

Step-by-step solution:

We rewrite the integral as $\int \frac{4x^2 + x + 1}{4x^3 + x} dx = \int \frac{4x^2 + 1}{4x^3 + x} dx + \int \frac{x}{4x^3 + x}$

dx. But $\int \frac{4x^2 + 1}{x(4x^2 + 1)} dx = \int \frac{dx}{x} = \ln x$ and $\int \frac{x}{x(4x^2 + 1)} dx = \int \frac{dx}{4x^2 + 1}$.

Now, let $2x = z \Rightarrow dx = \frac{dz}{2}$ and rewrite the new integral in terms

of the variable z to get $\int \frac{dx}{4x^2 + 1} = \frac{1}{2} \int \frac{1}{1 + z^2} dz = \frac{1}{2} \tan^{-1} z$. After col-

lecting both integrals results and write them in terms of the original

variable x we get $\int \frac{4x^2 + x + 1}{4x^3 + x} dx = \ln x + \frac{1}{2} \tan^{-1}(2x)$.

INTEGRAL 11

Problem

$$\int \frac{x^4 + 1}{x^2 + 2} \, dx$$

Solution:

$$\frac{x^3}{3} - 2x + \frac{5}{\sqrt{2}} \tan^{-1}\left(\frac{x}{\sqrt{2}}\right) + constant$$

Techniques used:

Change of variables, Partial fractions

Step-by-step solution:

We perform the division operation for the integrand to get $\frac{x^4 + 1}{x^2 + 2} = x^2 - 2 + \frac{5}{x^2 + 2}$. Therefore, the integral reads $\int \frac{x^4 + 1}{x^2 + 2} dx =$ $\int (x^2 - 2) dx + 5 \int \frac{1}{x^2 + 2} dx = \frac{x^3}{3} - 2x + 5 \int \frac{1}{x^2 + 2} dx$. But the new integral can be written as $\frac{5}{2} \int \frac{1}{1 + \left(\frac{x}{\sqrt{2}}\right)^2} dx$. Now, let $\frac{x}{\sqrt{2}} = z \Rightarrow dx$ $= \sqrt{2} dz$ and rewrite the last integral in terms of the variable z to get $\frac{5}{2} \int \frac{1}{1 + \left(\frac{x}{\sqrt{2}}\right)^2} dx = \frac{5\sqrt{2}}{2} \int \frac{1}{1 + z^2} dz = \frac{5\sqrt{2}}{2} \tan^{-1} z$. After collecting both integrals results and write them in terms of the original variable x we get $\int \frac{x^4 + 1}{x^2 + 2} dx = \frac{x^3}{3} - 2x + \frac{5}{\sqrt{2}} \tan^{-1}\left(\frac{x}{\sqrt{2}}\right)$.

INTEGRAL 12

Problem

$$\int \frac{1}{5 + 4\cos x}\, dx$$

Solution:

$$\frac{2}{3}\tan^{-1}\left[\frac{1}{3}\tan\left(\frac{x}{2}\right)\right] + constant$$

Techniques used:

Change of variables, Trigonometric identities

Step-by-step solution:

We rewrite the integral in terms of half-angle to have $\int \dfrac{1}{5 + 4\cos x}\, dx$

$$= \int \frac{1}{5 + 4\left(\cos^2\left(\dfrac{x}{2}\right) - \sin^2\left(\dfrac{x}{2}\right)\right)}\, dx. \text{ Let } \tan\frac{x}{2} = z \Rightarrow dz = \frac{1}{2}\frac{dx}{\cos^2\left(\dfrac{x}{2}\right)}.$$

Therefore, $\cos^2\left(\dfrac{x}{2}\right) = \dfrac{1}{1 + z^2}$, and $\sin^2\left(\dfrac{x}{2}\right) = \dfrac{z^2}{1 + z^2}$. Rewriting the integral in terms of the variable z, we get, after some simplifications,

$$\int \frac{1}{5 + 4\left(\cos^2\left(\dfrac{x}{2}\right) - \sin^2\left(\dfrac{x}{2}\right)\right)}\, dx = \int \frac{2}{9 + z^2}\, dz = \frac{2}{9}\int \frac{1}{1 + \left(\dfrac{z}{3}\right)^2}. \text{ Now let}$$

$\dfrac{z}{3} = u \Rightarrow dz = 3du$ and rewrite the integral in terms of the variable u. Therefore, we have $\dfrac{2}{3}\int \dfrac{du}{1 + u^2} = \dfrac{2}{3}\tan^{-1} u$. After rewriting this integral in terms of the original variable x, we get

$$\int \frac{1}{5 + 4\cos x}\, dx = \frac{2}{3}\tan^{-1}\left[\frac{1}{3}\tan\left(\frac{x}{2}\right)\right] \text{ as shown in the answer.}$$

INTEGRAL 13

Problem

$$\int \sec^4 x \, dx$$

Solution:

$$\frac{1}{3}\tan^3 x + \tan x + constant$$

Techniques used:

Integration by parts, Trigonometric identities

Step-by-step solution:

Rewrite the integral using trigonometric identities as $\int \sec^4 x \, dx =$

$\int \dfrac{dx}{\cos^4 x} = \int \dfrac{\overbrace{\sin^2 x + \cos^2 x}^{=1}}{\cos^4 x} dx$. Now, manipulate the integrand by

writing is as follows, $\int \dfrac{1}{\cos^4 x} dx = \int \dfrac{\overbrace{\sin^2 x + \cos^2 x}^{=1}}{\cos^4 x} dx = \int \dfrac{\sin^2 x}{\cos^4 x} dx +$

$\int \dfrac{dx}{\cos^2 x}$. But $\int \dfrac{dx}{\cos^2 x} = \tan x$, since $\dfrac{d}{dx}\tan x = \dfrac{1}{\cos^2 x}$.

For the other integral, we write $\int \dfrac{\sin^2 x}{\cos^4 x} dx = \int \dfrac{\sin^2 x}{\cos^2 x}\left(\dfrac{1}{\cos^2 x}\right) dx =$

$\int \tan^2 x \left(\dfrac{d}{dx}\tan x\right) = \dfrac{1}{3}\tan^3 x$. After collecting all terms, we get the

answer as $\int \sec^4 x \, dx = \dfrac{1}{3}\tan^3 x + \tan x$.

INTEGRAL 14

Problem:

$$\int \frac{1}{\sqrt{\left(1+x^2\right)^5}}\, dx$$

Solution:

$$\frac{3x+5x^3+2x^5}{3\sqrt{\left(1+x^2\right)^5}} + constant$$

Techniques used:

change of variables, Integration by parts, trigonometric identities

Step-by-step solution:

Let $x = \tan\alpha \Rightarrow dx = \dfrac{d\alpha}{\cos^2\alpha}, 1+x^2 = 1+\tan^2\alpha = \dfrac{1}{\cos^2\alpha}$. Now, rewriting the integral in terms of the variable α gives $\int \dfrac{1}{\sqrt{\left(1+x^2\right)^5}}\, dx = \int \dfrac{1}{\sqrt{\cos^{-10}\alpha}}\dfrac{d\alpha}{\cos^2\alpha} = \int \cos^3\alpha\, d\alpha$. Now we can write $\int \cos^3\alpha\, d\alpha = \int \cos^2\alpha\cos\alpha\, d\alpha = \int \left(1-\sin^2\alpha\right)\cos\alpha\, d\alpha = \int \cos\alpha\, d\alpha$ $-\int \sin^2\alpha\cos\alpha\, d\alpha$. But $\int \cos\alpha\, d\alpha = \sin\alpha$ and $-\int \sin^2\alpha\cos\alpha\, d\alpha =$ $-\int \sin^2\alpha\left(\dfrac{d}{d\alpha}\sin\alpha\right) = -\dfrac{1}{3}\sin^3\alpha$. Collecting all related terms gives the answer as $\int \cos^3\alpha\, d\alpha = \sin\alpha - \dfrac{1}{3}\sin^3\alpha$. Rewriting the result in terms of the original variable x, having $\alpha = \tan^{-1}x$, gives $\int \dfrac{1}{\sqrt{\left(1+x^2\right)^5}}\, dx = \sin\left(\tan^{-1}x\right) - \dfrac{1}{3}\sin^3\left(\tan^{-1}x\right)$. An alternative form of the answer, in terms of polynomial functions of x, can be obtained by substituting $\sin\alpha = \dfrac{x}{\sqrt{1+x^2}}$ into the expression $\sin\alpha - \dfrac{1}{3}\sin^3\alpha$ to get $\dfrac{3x+5x^3+2x^5}{3\sqrt{\left(1+x^2\right)^5}}$.

INTEGRAL 15

Problem:

$$\int \sqrt{\tan x}\, dx$$

Solution:

$$\frac{\sqrt{2}}{2}\left[\tan^{-1}\left(\sqrt{2\tan x}+1\right) + \tan^{-1}\left(\sqrt{2\tan x}-1\right) + \ln\sqrt{\frac{1+\tan x-\sqrt{2\tan x}}{1+\tan x+\sqrt{2\tan x}}}\right] + constant$$

Techniques used:

change of variables, partial fractions, trigonometric identities

Step-by-step solution:

Let $\tan x = z \Rightarrow \dfrac{dx}{\cos^2 x} = dz \Rightarrow dx = \dfrac{dz}{1+z^2}$. We rewrite the integral in terms of the variable z, to get $\int\sqrt{\tan x}\,dx = \int\dfrac{\sqrt{z}}{1+z^2}\,dz$. Now, let $\sqrt{z}=u \Rightarrow dz = 2u\,du$ and the latter integral can be written as $\int\dfrac{\sqrt{z}}{1+z^2}\,dz = 2\int\dfrac{u^2}{1+u^4}\,du$. The denominator of the integrand can be written as $1+u^4 = \left(1+u^2\right)^2 - 2u^2 = \left(1+u^2+\sqrt{2}u\right)\left(1+u^2-\sqrt{2}u\right)$. Therefore, using the partial fractions technique, we can write the integral as $2\int\dfrac{u^2}{1+u^4}\,du = 2\int\dfrac{u^2}{\left(1+u^2+\sqrt{2}u\right)\left(1+u^2-\sqrt{2}u\right)}\,du = \dfrac{1}{\sqrt{2}}$

$\int\dfrac{u}{1+u^2-\sqrt{2}u}\,du - \dfrac{1}{\sqrt{2}}\int\dfrac{u}{1+u^2+\sqrt{2}u}\,du$. Now we have two new integrals to calculate. Working on the latter, we can write the denominator of the integrand as $1+u^2+\sqrt{2}u=\left(u+\dfrac{\sqrt{2}}{2}\right)^2+$

$\dfrac{1}{2}=\dfrac{1}{2}\left[1+\left(\sqrt{2}u+1\right)^2\right]$. Now let $\sqrt{2}u+1=y\Rightarrow du=\dfrac{dy}{\sqrt{2}}$,

and write the latter integral in terms of variable y as

$$-\dfrac{1}{\sqrt{2}}\int\dfrac{u}{1+u^2+\sqrt{2}u}du=-\dfrac{2}{\sqrt{2}}\int\dfrac{(y-1)/\sqrt{2}}{1+y^2}\dfrac{dy}{\sqrt{2}}=\dfrac{1}{\sqrt{2}}\dfrac{1}{\sqrt{2}}\int\dfrac{1-y}{1+y^2}dy.$$

But $\dfrac{1}{\sqrt{2}}\int\dfrac{1-y}{1+y^2}dy=\dfrac{1}{\sqrt{2}}\int\dfrac{1}{1+y^2}dy-\dfrac{1}{2\sqrt{2}}\int\dfrac{2y}{1+y^2}dy$. The former

integral reads $\dfrac{1}{\sqrt{2}}\int\dfrac{1}{1+y^2}dy=\dfrac{1}{\sqrt{2}}\tan^{-1}y$ and the latter $-\dfrac{1}{2\sqrt{2}}$

$\int\dfrac{2y}{1+y^2}dy=-\dfrac{1}{2\sqrt{2}}\ln\left(1+y^2\right)$. Writing the results in terms of varia-

ble x, gives $\dfrac{1}{\sqrt{2}}\tan^{-1}y-\dfrac{1}{2\sqrt{2}}\ln\left(1+y^2\right)=\dfrac{1}{\sqrt{2}}\tan^{-1}\left(\sqrt{2}u+1\right)-\dfrac{1}{2\sqrt{2}}$

$\ln\left(1+\left(\sqrt{2}u+1\right)^2\right)=\dfrac{1}{\sqrt{2}}\tan^{-1}\left(\sqrt{2z}+1\right)-\dfrac{1}{2\sqrt{2}}\ln\left(1+\left(\sqrt{2z}+1\right)^2\right)=$

$\dfrac{1}{\sqrt{2}}\tan^{-1}\left(\sqrt{2\tan x}+1\right)-\dfrac{1}{2\sqrt{2}}\ln\left(1+\left(\sqrt{2\tan x}+1\right)^2\right)$.

Similarly, the remaining integral can be worked out and

reads as $\dfrac{1}{\sqrt{2}}\int\dfrac{u}{1+u^2-\sqrt{2}u}du=\dfrac{1}{\sqrt{2}}\tan^{-1}\left(\sqrt{2\tan x}-1\right)+\dfrac{1}{2\sqrt{2}}\ln\Big(1+$

$\left(\sqrt{2\tan x}-1\right)^2\Big)$. Therefore, solution reads as $\displaystyle\int\sqrt{\tan x}dx=\dfrac{\sqrt{2}}{2}\Big[\tan^{-1}$

$\left(\sqrt{2\tan x}+1\right)+\tan^{-1}\left(\sqrt{2\tan x}-1\right)\Big]+\dfrac{\sqrt{2}}{4}\ln\left[\dfrac{1+\left(\sqrt{2\tan x}-1\right)^2}{1+\left(\sqrt{2\tan x}+1\right)^2}\right]$

$$\left[\frac{1+\left(\sqrt{2\tan x}-1\right)^2}{1+\left(\sqrt{2\tan x}+1\right)^2}\right].$$ The last term simplifies to

$$\frac{\sqrt{2}}{4}\ln\left[\frac{1+\left(\sqrt{2\tan x}-1\right)^2}{1+\left(\sqrt{2\tan x}+1\right)^2}\right]=\frac{\sqrt{2}}{4}\ln\left[\frac{1+2\tan x+1-2\sqrt{2\tan x}}{1+2\tan x+1+2\sqrt{2\tan x}}\right]=\frac{\sqrt{2}}{4}$$

$$\ln 2+\frac{\sqrt{2}}{4}\ln\left[\frac{1+\tan x-\sqrt{2\tan x}}{1+\tan x+\sqrt{2\tan x}}\right].$$ Simplified solution reads as $\dfrac{\sqrt{2}}{2}$

$$\left[\tan^{-1}\left(\sqrt{2\tan x}+1\right)+\tan^{-1}\left(\sqrt{2\tan x}-1\right)+\ln\sqrt{\left(\frac{1+\tan x-\sqrt{2\tan x}}{1+\tan x+\sqrt{2\tan x}}\right)}\right]$$

$$+\frac{\sqrt{2}}{4}\ln 2.$$

INTEGRAL 16

Problem:

$$\int xe^x \sin x \, dx$$

Solution:

$$\frac{e^x}{2}\left[x\sin x + (1-x)\cos x\right] + constant$$

Techniques used:

Integration by parts, trigonometric identities

Step-by-step solution:

Knowing that $\frac{d}{dx}(xe^x) = e^x(x+1)$ and $\int xe^x dx = e^x(x-1)$, we apply the integration by parts technique to the integral to get $\int \underset{f}{\underbrace{xe^x}} \underset{dg}{\underbrace{\sin x dx}} = -xe^x \cos x + \int e^x(x+1)\cos x dx = -xe^x \cos x + \int xe^x$

$\cos x dx + \int e^x \cos x dx$. But $\int \underset{f}{\underbrace{xe^x}} \underset{dg}{\underbrace{\cos x dx}} = xe^x \sin x - \int e^x(x+1)\sin x dx$

and Therefore, after substitution, we get $\int xe^x \sin x dx = -xe^x \cos x$

$+ xe^x \sin x - \int e^x(x+1)\sin x dx + \int e^x \cos x dx$. Or, after rearranging

the involved terms, we get $\int xe^x \sin x dx = \frac{1}{2}\left[-xe^x \cos x + xe^x \sin x - \right.$

$\left. \int e^x \sin x dx + \int e^x \cos x dx\right]$. But $-\int e^x \sin x dx = e^x \cos x - \int e^x \cos x dx$.

Therefore, after substitution, we get $\int xe^x \sin x dx = \frac{1}{2}\left[-xe^x \cos x + \right.$

$xe^x \sin x + e^x \cos x - \int e^x \cos x dx + \int e^x \cos x dx\right] = \frac{1}{2}\left[-xe^x \cos x + xe^x \right.$

$\left. \sin x + e^x \cos x\right]$. Or $\int xe^x \sin x dx = \frac{e^x}{2}\left[x\sin x + (1-x)\cos x\right]$.

INTEGRAL 17

Problem:

$$\int \frac{\ln^3 x}{x^3}\,dx$$

Solution:

$$\frac{1}{8x^2}\left[4\ln^3 x + 6\ln^2 x + 6\ln x + 3\right] + constant$$

Techniques used:

Integration by parts

Step-by-step solution:

Rewrite the integrand and apply the integration by parts. Therefore, we get $\int \frac{\ln^3 x}{x^3}\,dx = \int \underbrace{\ln^3 x}_{f}\underbrace{\left(\frac{dx}{x^3}\right)}_{dg} = \frac{-1}{2x^2}\ln^3 x + \frac{3}{2}\int \frac{\ln^2 x}{x^3}\,dx$. Applying the integration by parts successively twice again to the last integral, gives

$$\frac{3}{2}\int \frac{\ln^2 x}{x^3}\,dx = \frac{3}{2}\left[\frac{-1}{2x^2}\ln^2 x + \int \frac{\ln x}{x^3}\,dx\right] = \frac{3}{2}\left[\frac{-1}{2x^2}\ln^2 x - \frac{1}{2x^2}\ln x + \frac{1}{2}\right]$$

$$\int \frac{1}{x^3}\,dx\Big] = \frac{3}{2}\left[\frac{-1}{2x^2}\ln^2 x - \frac{1}{2x^2}\ln x - \frac{1}{4x^2}\right].$$ After collecting all terms, we

receive the answer as $\int \frac{\ln^3 x}{x^3}\,dx = \frac{-3}{8x^2}\left[1 + 2\ln x\left(1 + \ln x + \frac{2}{3}\ln^2 x\right)\right].$

INTEGRAL 18

Problem:

$$\int \ln\left(1+\sqrt{x}\right) dx$$

Solution:

$$(x-1)\ln\left(1+\sqrt{x}\right)+\sqrt{x}-\frac{x}{2}+constant$$

Techniques used:

Change of variables, Integration by parts, Partial fractions

Step-by-step solution:

We use the integration by parts technique and consider $\int dx = x$.

Therefore, we can write $\int \ln\left(1+\sqrt{x}\right) dx = x\ln\left(1+\sqrt{x}\right) - \int x \frac{\frac{1}{2\sqrt{x}}}{1+\sqrt{x}} dx$

$= x\ln\left(1+\sqrt{x}\right) - \frac{1}{2}\int \frac{x}{x+\sqrt{x}} dx$. Now, let $\sqrt{x} = z \Rightarrow dx = 2zdz$ and write the new integral in terms of variable z as

$-\frac{1}{2}\int \frac{x}{x+\sqrt{x}} dx = -\int \frac{z^3}{z^2+z} dz = -\int \frac{z^2}{z+1} dz$. Using partial fractions,

we can write $-\int \frac{z^2}{z+1} dz = -\int \left(z-1+\frac{1}{z+1}\right) dz$. Performing the

integration, gives $-\int \left(z-1+\frac{1}{z+1}\right) dz = -\frac{1}{2}z^2 + z - \ln(1+z)$. After

collecting all terms and write them in term of the original variable

x, we get the answer as $(x-1)\ln\left(1+\sqrt{x}\right)-\frac{1}{2}x+\sqrt{x}$.

INTEGRAL 19

Problem:

$$\int \ln\left(1 + \sqrt{1 + \frac{1}{x^2}}\right) dx$$

Solution:

$$x\left[\ln\left(1 + \sqrt{1 + \frac{1}{x^2}}\right) + 1 - \sqrt{1 + \frac{1}{x^2}}\right] + constant$$

Techniques used:

Change of variables, Integration by parts

Step-by-step solution:

We use the integration by parts technique along with considering $\int dx = x$. Therefore, we can write $\int \underbrace{\ln\left(1 + \sqrt{1 + \frac{1}{x^2}}\right)}_{f}$

$$dx = x \ln\left(1 + \sqrt{1 + \frac{1}{x^2}}\right) - \int x \frac{\left(1 + \sqrt{1 + \frac{1}{x^2}}\right)'}{1 + \sqrt{1 + \frac{1}{x^2}}} dx,$$

with the prime symbol indicating differentiation. Therefore,

$$-\int x \frac{\left(1 + \sqrt{1 + \frac{1}{x^2}}\right)'}{1 + \sqrt{1 + \frac{1}{x^2}}} dx = \int \frac{dx}{x^2\sqrt{1 + \frac{1}{x^2}} + x^2 + 1}.$$ After simplifying the

integrand, we get $\int \dfrac{dx}{x^2\sqrt{1 + \frac{1}{x^2}} + x^2 + 1} = \int \dfrac{dx}{1 + x^2 + x\sqrt{1 + x^2}}.$ Let

$\sqrt{1 + x^2} = z \Rightarrow dx = \dfrac{z dz}{\sqrt{z^2 - 1}}.$ Therefor, the latter integral in

terms of variable z can be written as $\int \dfrac{1}{z\sqrt{z^2 - 1} + z^2 - 1} dz$

$$= \int \frac{1}{\sqrt{z^2-1}\left(z+\sqrt{z^2-1}\right)}dz. \quad \text{Now,} \quad \text{let} \quad z+\sqrt{z^2-1}=u \Rightarrow dz=$$

$$\frac{du}{\dfrac{z}{\sqrt{z^2-1}}+1}. \text{ Rewriting the latter integral in terms of the variable}$$

u, we have $\displaystyle \int \frac{1}{\sqrt{z^2-1}\left(z+\sqrt{z^2-1}\right)}dz = \int \frac{du}{u\left(\dfrac{z}{\sqrt{z^2-1}}+1\right)\sqrt{z^2-1}} =$

$$\underbrace{\phantom{u\left(\dfrac{z}{\sqrt{z^2-1}}+1\right)\sqrt{z^2-1}}}_{u}$$

$= \displaystyle \int \frac{du}{u^2} = -\frac{1}{u}$. After back substituting for u in terms of z and x we get

the results as $-\dfrac{1}{u} = -\dfrac{1}{z+\sqrt{z^2-1}} = -\dfrac{1}{\sqrt{1+x^2}+x}$. Multiplying

the last expression by its conjugates ratio gives,

$-\dfrac{1}{\sqrt{1+x^2}+x} \dfrac{\sqrt{1+x^2}-x}{\sqrt{1+x^2}-x} = \dfrac{x-\sqrt{1+x^2}}{1+x^2-x^2} = x-\sqrt{1+x^2}$. After collect-

ing all related terms, we get the final answer as

$\displaystyle \int \ln\left(1+\sqrt{1+\dfrac{1}{x^2}}\right)dx = x\ln\left(1+\sqrt{1+\dfrac{1}{x^2}}\right)+x-\sqrt{1+x^2}$. Please note

that $\sqrt{1+x^2} = x\sqrt{1+\dfrac{1}{x^2}}$.

INTEGRAL 20

Problem:

$$\int \operatorname{csch}^{-1} x \, dx$$

Solution:

$$x \operatorname{csch}^{-1} x + \sinh^{-1} x + constant$$

Techniques used:

Change of variables, Integration by parts

Step-by-step solution:

Let $\operatorname{csch}^{-1} x = \alpha \Rightarrow \operatorname{csch} \alpha = \dfrac{1}{\sinh \alpha} = x.$ Therefore, $\sinh \alpha = \dfrac{1}{x}$ and $\cosh \alpha \, d\alpha = -\dfrac{dx}{x^2}.$ Now we can write the integral in terms of variable α, as $\int \operatorname{csch}^{-1} x \, dx = \int \sinh^{-1}\left(\dfrac{1}{x}\right) = -\int \alpha \left(\dfrac{\cosh \alpha}{\sinh^2 \alpha}\right) d\alpha.$ Using the integration by parts, we have $-\int \alpha \left(\dfrac{\cosh \alpha}{\sinh^2 \alpha}\right) d\alpha = \dfrac{\alpha}{\sinh \alpha} - \int \dfrac{d\alpha}{\sinh \alpha}.$ But writing the latter integral in terms of original variable x, we get $-\int \dfrac{d\alpha}{\sinh \alpha} = \int \dfrac{dx}{\cosh \alpha . x^2} \dfrac{1}{\sinh \alpha} = \int \dfrac{dx}{x} \cdot \dfrac{x}{\sqrt{1 + x^2}} = \int \dfrac{dx}{\sqrt{1 + x^2}} = \sinh^{-1} x.$ Now collecting all related terms, we get the solution as $\dfrac{\alpha}{\sinh \alpha} + \sinh^{-1} x = x \operatorname{csch}^{-1} x + \sinh^{-1} x,$ for $x > 0.$

INTEGRAL 21

Problem:

$$\int \frac{1}{\sin x + \cos x}\, dx$$

Solution:

$$\frac{\sqrt{2}}{2} \ln \left[\frac{\tan\left(\dfrac{x}{2}\right) - 1 + \sqrt{2}}{\tan\left(\dfrac{x}{2}\right) - 1 - \sqrt{2}} \right] + constant$$

Techniques used:

Change of variables, Partial fractions, Trigonometric identities

Step-by-step solution:

Let $\tan\dfrac{x}{2} = z$, then we have $dx = 2dz / \left(1 + z^2\right)$. Using the trigo-

nometric identities, we have $\sin x = \dfrac{2\tan\left(\dfrac{x}{2}\right)}{1 + \tan^2\left(\dfrac{x}{2}\right)} = \dfrac{2z}{1 + z^2}$, and

$\cos x = \dfrac{1 - \tan^2\left(\dfrac{x}{2}\right)}{1 + \tan^2\left(\dfrac{x}{2}\right)} = \dfrac{1 - z^2}{1 + z^2}$. Therefore, the integral can be written

in terms of the variable z as $\displaystyle\int \frac{1}{\sin x + \cos x}\, dx = 2\int \frac{1}{1 + 2z - z^2}\, dz$.

But the denominator of the integrand can be written as

$1 + 2z - z^2 = -\left(z^2 - 2z - 1\right) = -\left[\left(z - 1\right)^2 - 2\right] = -\left(z - 1 - \sqrt{2}\right)\left(z - 1\right)$

$+ \sqrt{2}$. Now we rewrite the integral as $2\displaystyle\int \frac{1}{1 + 2z - z^2}\, dz =$

$$-2\int \frac{1}{\left(z-1-\sqrt{2}\right)\left(z-1+\sqrt{2}\right)}\,dz.$$ Using the partial fractions

technique, we can write $-2\int \frac{1}{\left(z-1-\sqrt{2}\right)\left(z-1+\sqrt{2}\right)}\,dz =$

$-\dfrac{\sqrt{2}}{2}\displaystyle\int \dfrac{dz}{z-1-\sqrt{2}} + \dfrac{\sqrt{2}}{2}\displaystyle\int \dfrac{dz}{z-1+\sqrt{2}}$. Finally, we can write the

solutions by integrating each term, as $-\dfrac{\sqrt{2}}{2}\ln\left(z-1-\sqrt{2}\right)+$

$\dfrac{\sqrt{2}}{2}\ln\left(z-1+\sqrt{2}\right) = \dfrac{\sqrt{2}}{2}\ln\left(\dfrac{z-1+\sqrt{2}}{z-1-\sqrt{2}}\right)$. Therefore, the answer in

terms of the original variable x is $\dfrac{\sqrt{2}}{2}\ln\left(\dfrac{z-1+\sqrt{2}}{z-1-\sqrt{2}}\right) = \dfrac{\sqrt{2}}{2}$

$\ln\left[\dfrac{\tan\left(\dfrac{x}{2}\right)-1+\sqrt{2}}{\tan\left(\dfrac{x}{2}\right)-1-\sqrt{2}}\right]$. Please note that the absolute value of the

argument of the logarithm should be considered.

INTEGRAL 22

Problem:

$$\int 2^{\sqrt{-x}}\, dx$$

Solution:

$$\frac{2^{1+\sqrt{-x}}}{\log^2 2}\left(1 - \sqrt{-x}\log 2\right) + constant$$

Techniques used:

Change of variables, Logarithm identities

Step-by-step solution:

Let $2^{\sqrt{-x}} = z \Rightarrow \log_2 z = \sqrt{-x} = i\sqrt{x}$. We can write $\log_2 z =$

$\dfrac{\log z}{\log 2} = i\sqrt{x}$, or $\dfrac{dz}{z\log 2} = \dfrac{i\,dx}{2\sqrt{x}}$. Therefore, $dx = \dfrac{2\sqrt{x}}{iz\log 2} = \dfrac{2\sqrt{-x}}{i^2 z\log 2} =$

$\dfrac{2(\log z/\log 2)}{-z\log 2} = -\dfrac{2\log z}{z\log^2 2}$. Rewriting the integral in terms of the var-

iable z, we get $\int 2^{\sqrt{-x}}\, dx = -\dfrac{2}{\log^2 2}\int \log z\, dz$. But $\int \log z\, dz = z\log z - z$,

and after rewriting this expression in terms of the original x varaiable we

get $\int 2^{\sqrt{-x}}\, dx = -\dfrac{2}{\log^2 2}\left(2^{\sqrt{-x}}\log 2^{\sqrt{-x}} - 2^{\sqrt{-x}}\right) = \dfrac{2^{1+\sqrt{-x}}}{\log^2 2}\left(1 - \sqrt{-x}\log 2\right).$

INTEGRAL 23

Problem:

$$\int \ln\left(1 + \sqrt{1 + x^2}\right) dx$$

Solution:

$$x \ln\left(1 + \sqrt{1 + x^2}\right) + \sinh^{-1} x - x + constant$$

Techniques used:

Change of variables, Trigonometric identities

Step-by-step solution:

Using the integration by parts technique, we can write the integral as $\underbrace{\int \ln\left(1 + \sqrt{1 + x^2}\right)}_{f} \underbrace{dx}_{dg} = x \ln\left(1 + \sqrt{1 + x^2}\right) - \int x \, f' dx$.

But $f' = \dfrac{d}{dx} \ln\left(1 + \sqrt{1 + x^2}\right) = \dfrac{x}{1 + x^2 + \sqrt{1 + x^2}}$. Therefore, the latter

integral reads $\int x f' dx = -\int \dfrac{x^2}{1 + x^2 + \sqrt{1 + x^2}} dx$. Let $\sqrt{1 + x^2} = z \Rightarrow dx$

$= z \dfrac{dz}{x} = \dfrac{z dz}{\sqrt{z^2 - 1}}$ and write the latter integral in terms of variable

z, as $-\int \dfrac{x^2}{1 + x^2 + \sqrt{1 + x^2}} dx = -\int \dfrac{z^2 - 1}{z^2 + z} \dfrac{z dz}{\sqrt{z^2 - 1}} = -\int \dfrac{\sqrt{z^2 - 1}}{z + 1} dz$. Now,

let $z = \cosh u \Rightarrow dz = \sinh u \, du$ and write the latter integral in terms

of variable u as $-\int \dfrac{\sqrt{z^2 - 1}}{z + 1} dz = -\int \dfrac{\sinh^2 u}{1 + \cosh u} du = -\int \dfrac{\cosh^2 u - 1}{1 + \cosh u} du$

$= \int(-\cosh u + 1)du = u - \sinh u$. This expression in terms of the original variable x reads $u - \sinh u = \cosh^{-1} z - \sqrt{z^2 - 1}$ $= \cosh^{-1} \sqrt{1 + x^2} - \sqrt{1 + x^2 - 1} = \cosh^{-1} \sqrt{1 + x^2} - x$, for $x > 0$. Collecting all related terms, we get the solution as $x \ln\left(1 + \sqrt{1 + x^2}\right) + \cosh^{-1} \sqrt{1 + x^2} - x$. But $\cosh^{-1} \sqrt{1 + x^2} = \sinh^{-1} x$.

Hence, the solution simplifies to $x \ln\left(1 + \sqrt{1 + x^2}\right) + \sinh^{-1} x - x$.

INTEGRAL 24

Problem:

$$\int \sin^6 x \cos^5 x dx$$

Solution:

$$\frac{8}{693}\sin^{11} x + \frac{4}{63}\sin^9 x \cos^2 x + \frac{1}{7}\sin^7 x \cos^4 x + constant$$

Techniques used:

Integration by parts

Step-by-step solution:

Rewrite the integral as $\int \sin^6 x \cos^5 x dx = \int (\sin^6 x \cos x)\cos^4 x dx$. Using the integration by parts, we get $\int \underbrace{\left(\sin^6 x \cos x \right)}_{dg} \underbrace{\cos^4 x}_{f} dx =$

$\frac{1}{7}\sin^7 x \cos^4 x + \frac{4}{7}\int \sin^8 x \cos^3 x dx$ Repeating the similar operation

on the latter integral, we get $\int \sin^8 x \cos^3 x dx = \int \sin^8 x \cos x \cos^2 x dx =$

$\frac{1}{9}\sin^9 x \cos^2 x + \frac{2}{9}\int \sin^{10} x \cos x dx$. But the latter integral reads

$\int \sin^{10} x \cos x dx = \frac{1}{11}\sin^{11} x$. Collecting all related results, we get

the solution as $\frac{8}{693}\sin^{11} x + \frac{4}{63}\sin^9 x \cos^2 x + \frac{1}{7}\sin^7 x \cos^4 x$.

INTEGRAL 25

Problem:

$$\int x^m \ln^n x\, dx$$

m and n are positive integers.

Solution:

$$\frac{x^{m+1}\ln^n x}{m+1} - \frac{n}{m+1}\int x^m \ln^{n-1} x\, dx + constant$$

For $m = n = 3$

$$\frac{x^4}{128}\left(32\ln^3 x - 24\ln^2 x + 12\ln x - 3\right)$$

Techniques used:

Integration by parts, Recursive relation

Step-by-step solution:

The solution provides a recursive relation applicable to integer values of m and n. We work out the general solution and apply it to a numerical example.

Applying the integration by parts technique, we can write $\int x^m \ln^n x\, dx = \frac{x^{m+1}}{m+1}\ln^n x - \frac{n}{m+1}\int x^m \ln^{n-1} x\, dx$. For example, for $m = n = 3$, we have $\int x^3 \ln^3 x\, dx = \frac{x^4}{4}\ln^3 x - \frac{3}{4}\int x^3 \ln^2 x\, dx$. But, using the general relation again for $m = 3, n = 2$, we get $\int x^3 \ln^2 x\, dx = \frac{x^4}{4}\ln^2 x - \frac{2}{4}\int x^3 \ln x\, dx$. Using the general formula for the latter integral, for $m = 3, n = 1$, we get $\int x^3 \ln x\, dx = \frac{x^4}{4}\ln x - \frac{1}{4}\int x^3 dx,$

with $\int x^3 dx = \dfrac{x^4}{4}$. After collecting all related terms, we get

$$\int x^3 \ln^3 x\, dx = \frac{x^4}{4}\ln^3 x - \frac{3}{4}\left(\frac{x^4}{4}\ln^2 x - \frac{2}{4}\left(\frac{x^4}{4}\ln x - \frac{1}{4}\frac{x^4}{4}\right)\right).$$ Or, after

simplification, we get the final answer as $\int x^3 \ln^3 x\, dx = \dfrac{x^4}{128}$

$\left(32\ln^3 x - 24\ln^2 x + 12\ln x - 3\right).$

INTEGRAL 26

Problem:

$$\int e^x \sin^{-1}\left(e^x\right) dx$$

Solution:

$$e^x \sin^{-1}\left(e^x\right) + \sqrt{1 - e^{2x}} + constant$$

Techniques used:

Change of variables, Integration by parts

Step-by-step solution:

Let $e^x = z \Rightarrow dx = dz / z$. Rewriting the integral in terms of the variable z, we get $\int e^x \sin^{-1}\left(e^x\right) dx = \int \sin^{-1} z \, dz$. Now, let $\sin^{-1} z = \alpha \Rightarrow \sin \alpha = z \Rightarrow \cos \alpha \, d\alpha = dz$. After back substituting, we get $\int \sin^{-1} z \, dz = \int \alpha \cos \alpha \, d\alpha$. Using the integration by parts technique, we get the answer, $\int \underset{f}{\underbrace{\alpha}} \underset{dg}{\underbrace{\cos \alpha \, d\alpha}} = \alpha \sin \alpha - \int \sin \alpha \, d\alpha = \alpha \sin \alpha + \cos \alpha$.

Rewriting the results in terms of variables z, and then x, we get $\alpha \sin \alpha + \cos \alpha = z \sin^{-1} z + \sqrt{1 - z^2} = e^x \sin^{-1}\left(e^x\right) + \sqrt{1 - e^{2x}}$.

INTEGRAL 27

Problem:

$$\int x\sqrt{\frac{1-x^2}{1+x^2}}\,dx$$

Solution:

$$\frac{1}{2}\left(\sqrt{1-x^4}-2\tan^{-1}\sqrt{\frac{1-x^2}{1+x^2}}\right)+constant$$

Techniques used:

Change of variables, Partial fractions, Trigonometric identities

Step-by-step solution:

Let $x=\tan(z/2)\Rightarrow dx=\dfrac{dz}{2\cos^2(z/2)}$. Therefore, the integral

can be written in terms of the variable z as $\int x\sqrt{\dfrac{1-x^2}{1+x^2}}\,dx=$

$\dfrac{1}{2}\displaystyle\int\dfrac{\tan(z/2)}{\cos^2(z/2)}\sqrt{\dfrac{1-\tan^2(z/2)}{1+\tan^2(z/2)}}\,dz.$ But $\sqrt{\dfrac{1-\tan^2(z/2)}{1+\tan^2(z/2)}}=\sqrt{\cos z},$

hence we get $\dfrac{1}{2}\displaystyle\int\dfrac{\tan(z/2)}{\cos^2(z/2)}\sqrt{\dfrac{1-\tan^2(z/2)}{1+\tan^2(z/2)}}dz=\dfrac{1}{2}\displaystyle\int\dfrac{\sin(z/2)}{\cos^3(z/2)}$

$\sqrt{\cos z}\,dz.$ Now, after multiplying the integrand by $\dfrac{\cos(z/2)}{\cos(z/2)}$, we

have $\dfrac{1}{2}\displaystyle\int\dfrac{\sin(z/2)\cos(z/2)}{\cos^4(z/2)}\sqrt{\cos z}\,dz=\dfrac{1}{4}\displaystyle\int\dfrac{\sin z}{\cos^4(z/2)}\sqrt{\cos z}\,dz.$ Let

$\cos z=u\Rightarrow-\sin z\ dz=du$ and $\dfrac{\sqrt{\cos z}}{\cos^4(z/2)}=\dfrac{4\sqrt{u}}{(1+u)^2}.$ Therefore, we

have $\dfrac{1}{4}\displaystyle\int\dfrac{\sin z}{\cos^4(z/2)}\sqrt{\cos z}\,dz=-\displaystyle\int\dfrac{\sqrt{u}}{(1+u)^2}\,du.$ Now let $\sqrt{u}=t\Rightarrow du$

$=2tdt$ and writing the latter integral in terms of the variable t, we get $-2\int \dfrac{t^2}{\left(1+t^2\right)^2}\,dt$. After using the partial fractions techniques,

we can write $-2\int \dfrac{t^2}{\left(1+t^2\right)^2}\,dt = 2\left(\int \dfrac{1}{\left(1+t^2\right)^2}\,dt - \int \dfrac{1}{\left(1+t^2\right)}\,dt\right)$.

The latter integral reads $-2\int \dfrac{1}{\left(1+t^2\right)}\,dt = -2\tan^{-1} t$. But for calcu-

lating the integral $2\int \dfrac{1}{\left(1+t^2\right)^2}\,dt$ we let $t = \tan y \Rightarrow dt = dy\,/\cos^2 y$

and $\left(1+t^2\right)^2 = \left(1+\tan^2 y\right)^2 = \left(\dfrac{\cos^2 y + \sin^2 y}{\cos^2 y}\right)^2 = \dfrac{1}{\cos^4 y}$. After

substation, we get $2\int \dfrac{1}{\left(1+t^2\right)^2}\,dt = 2\int \dfrac{\cos^4 y}{\cos^2 y}\,dy = 2\int \cos^2 y\,dy$. But,

using trigonometric identity $\cos^2 y = \dfrac{1+\cos 2y}{2}$, we get $2\int \cos^2 y\,dy$

$= \int \left(1+\cos 2y\right)dy = y + \dfrac{1}{2}\sin 2y$. Now, after a series of

back substitutions and collecting related terms, we get

$\sqrt{\dfrac{\cos\left(2\tan^{-1} x\right)}{1+\cos\left(2\tan^{-1} x\right)}} - \tan^{-1}\sqrt{\cos\left(2\tan^{-1} x\right)}$. This expression simpli-

fies to $\dfrac{1}{2}\left(\sqrt{1-x^4} - 2\tan^{-1}\sqrt{\dfrac{1-x^2}{1+x^2}}\right)$.

INTEGRAL 28

Problem:

$$\int x\sqrt{x}\tan^{-1}\sqrt{x}\,dx$$

Solution:

$$\frac{1}{10}\left[4x^2\sqrt{x}\tan^{-1}\sqrt{x}-2\ln(1+x)-x(x-2)\right]+constant$$

Techniques used:

Change of variables, Integration by parts, Partial fractions

Step-by-step solution:

Let $\sqrt{x}=z\Rightarrow dx=2zdz$, rewriting the integral in terms of the variable z gives $\int x\sqrt{x}\tan^{-1}\sqrt{x}\,dx=2\int z^4\tan^{-1}z\,dz$. Using the integration by parts technique we get $2\int z^4\tan^{-1}z\,dz=2\left(\frac{1}{5}z^5\tan^{-1}z-\frac{1}{5}\int\frac{z^5}{1+z^2}\,dz\right)$.
Note that we used the derivative of inverse tangent function, or $\frac{d}{dz}\left(\tan^{-1}z\right)=\frac{1}{1+z^2}$. Now, we work out the latter integral by using the partial fractions technique to get
$$\frac{1}{5}\int\frac{z^5}{1+z^2}\,dz=\frac{1}{5}\int\left(z^3-z+\frac{z}{1+z^2}\right)dz=\frac{1}{5}\left(\frac{z^4}{4}-\frac{z^2}{2}\right)+\frac{1}{5}\int\frac{z}{1+z^2}\,dz.$$
But $\frac{1}{5}\int\frac{z}{1+z^2}\,dz=\frac{1}{10}\ln(1+z^2)$. Collecting all related terms, we get
$$2\int z^4\tan^{-1}z\,dz=2\left[\frac{1}{5}z^5\tan^{-1}z-\frac{1}{5}\left(\frac{z^4}{4}-\frac{z^2}{2}\right)-\frac{1}{10}\ln(1+z^2)\right].$$ After
simplification and rewriting this expression in terms of the original variable x, we get $\frac{1}{10}\left[4x^2\sqrt{x}\tan^{-1}\sqrt{x}-2\ln(1+x)-x(x-2)\right]$.

INTEGRAL 29

Problem:

$$\int \frac{\tan^{-1} x}{(x-1)^3} dx$$

Solution:

$$\frac{1}{8}\left[\ln\frac{\left(x^2+1\right)}{(x-1)^2} - \frac{4}{(x-1)^2}\tan^{-1}x - \frac{2}{x-1} \right] + constant$$

Techniques used:

Integration by parts, Partial fractions

Step-by-step solution:

Using integration by parts technique we get, $\int \underbrace{\tan^{-1}x}_{f}\underbrace{\frac{dx}{(x-1)^3}}_{dg} =$

$-\frac{1}{2(x-1)^2}\tan^{-1}x + \frac{1}{2}\int \frac{1}{(x-1)^2}\cdot\frac{1}{1+x^2}dx.$ But using the par-

tial fraction technique we can write the latter integral as

$\frac{1}{2}\int \frac{1}{(x-1)^2}\frac{1}{1+x^2}dx = \frac{1}{4}\int \frac{x}{1+x^2}dx - \frac{1}{4}\int \frac{1}{x-1}dx + \frac{1}{4}\int \frac{1}{(x-1)^2}dx.$

But $\frac{1}{4}\int \frac{x}{1+x^2}dx = \frac{1}{8}\ln\left(1+x^2\right),$ and $\frac{1}{4}\int \frac{1}{x-1}dx = \frac{1}{4}\ln(x-1).$ For

the third integral in the last expression, we let $x-1=z \Rightarrow dx = dz.$

Therefore, $\frac{1}{4}\int \frac{1}{(x-1)^2}dx = \frac{1}{4}\int \frac{dz}{z^2} = -\frac{1}{4z} = -\frac{1}{4(x-1)}.$ Collecting

all the related terms we get the answer as $-\frac{1}{2(x-1)^2}\tan^{-1}x +$

$\frac{1}{8}\ln\left(1+x^2\right) - \frac{1}{4}\ln(x-1) - \frac{1}{4(x-1)}.$

INTEGRAL 30

Problem:

$$\int \frac{\sin^{-1} x}{x^2} dx$$

Solution:

$$\ln\left(\frac{x}{1 + \sqrt{1 - x^2}} \right) - \frac{\sin^{-1} x}{x} + constant$$

Techniques used:

Integration by parts, Change of variables, Partial fractions.

Step-by-step solution:

Using the integration by parts technique, we can write $\int \frac{\sin^{-1} x}{x^2} dx = \int \underbrace{\sin^{-1} x}_{f} \underbrace{\left(\frac{dx}{x^2} \right)}_{dg} = -\frac{\sin^{-1} x}{x} + \int \frac{1}{x\sqrt{1 - x^2}} dx$. Note

that $\frac{d}{dx} \sin^{-1} x = \frac{1}{\sqrt{1 - x^2}}$. Let $x = \cos z \Rightarrow dx = -\sin z \, dz$, then

$\int \frac{1}{x\sqrt{1 - x^2}} dx = -\int \frac{\sin z \, dz}{\cos z \sin z} = -\int \frac{dz}{\cos z} = -\int \sec z \, dz$. We calculated

the integral of $\sec z$ in the previous section (see Integral 5). Therefore,

we have $-\int \frac{dz}{\cos z} = -\ln\left(\frac{\cos(z/2) + \sin(z/2)}{\cos(z/2) - \sin(z/2)} \right)$. But we have

$x = \cos z = 2\cos^2(z/2) - 1 = 1 - 2\sin^2(z/2)$. Therefore, after some

manipulations, we get $\cos(z/2) = \sqrt{\frac{1 + x}{2}}$ and $\sin(z/2) = \sqrt{\frac{1 - x}{2}}$.

After back substituting, we get $-\displaystyle\int \frac{dz}{\cos z} = -\ln\left(\dfrac{\cos\left(\dfrac{z}{2}\right) + \sin\left(\dfrac{z}{2}\right)}{\cos\left(\dfrac{z}{2}\right) - \sin\left(\dfrac{z}{2}\right)} \right) =$

$-\ln\dfrac{\sqrt{1+x}+\sqrt{1-x}}{\sqrt{1+x}-\sqrt{1-x}} = -\ln\left(\dfrac{1+\sqrt{1-x^2}}{x} \right).$ Final   answer,   after

writing the results in terms of the original variable x, is

$\displaystyle\int \frac{\sin^{-1}x}{x^2}dx = -\frac{\sin^{-1}x}{x} - \ln\left(\frac{1+\sqrt{1-x^2}}{x} \right).$

INTEGRAL 31

Problem:

$$\int x \sec^{-1} x \, dx$$

Solution:

$$\frac{1}{2}\left(x^2 \sec^{-1} x - \sqrt{x^2 - 1}\right) + constant$$

Techniques used:

Integration by parts, Trigonometric identities

Step-by-step solution:

Using the trigonometric identity $\sec^{-1} x = \cos^{-1}(1/x)$ and $\frac{d}{dx}\left(\cos^{-1}(1/x)\right) = \frac{1}{x\sqrt{x^2-1}}$, we can rewrite the integral by applying the integration by parts technique as $\int x \sec^{-1} x$ $dx = \int \underbrace{\cos^{-1}\left(\frac{1}{x}\right)}_{f} \underbrace{\left(x dx\right)}_{dg} = \frac{1}{2}x^2 \cos^{-1}(1/x) - \frac{1}{2}\int \frac{x^2}{x\sqrt{x^2-1}} dx$. But $-\frac{1}{2}$

$\int \frac{x^2}{x\sqrt{x^2-1}} dx = -\frac{1}{4}\int \frac{2x}{\sqrt{x^2-1}} = -\frac{1}{2}\sqrt{x^2-1}$. Collecting the related

terms, we get the answer as $\int x \sec^{-1} x \, dx = \frac{1}{2}\left[x^2 \cos^{-1}(1/x)\right.$ $\left. - \sqrt{x^2-1}\right]$.

INTEGRAL 32

Problem:

$$\int \sec^{-1} \sqrt{x}\, dx$$

Solution:

$$\left(x\sec^{-1} \sqrt{x} - \sqrt{x-1} \right) + constant$$

Techniques used:

Integration by parts, Trigonometric identities

Step-by-step solution:

Using the trigonometric identity $\sec^{-1} \sqrt{x} = \cos^{-1}\left(1/\sqrt{x} \right)$ and $\dfrac{d}{dx}\left(\cos^{-1}\left(1/\sqrt{x} \right) \right) = \dfrac{1}{2x\sqrt{x-1}}$, we can rewrite the integral by applying the integration by parts technique as $\int \sec^{-1} \sqrt{x}\, dx =$

$$\underbrace{\int \cos^{-1}\left(\frac{1}{\sqrt{x}} \right)}_{f} \underbrace{\left(dx \right)}_{dg} = x\cos^{-1}\left(1/\sqrt{x} \right) - \frac{1}{2}\int \frac{1}{\sqrt{x-1}}\, dx = x\cos^{-1}\left(\frac{1}{\sqrt{x}} \right) -$$

$$\sqrt{x-1} = x\sec^{-1} \sqrt{x} - \sqrt{x-1}.$$

INTEGRAL 33

Problem:

$$\int x^2 \tan^{-1} x\, dx$$

Solution:

$$\frac{1}{6}\left[2x^3 \tan^{-1} x - x^2 + \ln\left(1 + x^2\right) \right] + constant$$

Techniques used:

Integration by parts, Partial fractions

Step-by-step solution:

Using the relation $\dfrac{d}{dx}\left(\tan^{-1} x\right) = \dfrac{1}{1+x^2}$, we can rewrite the integral by applying the integration by parts technique as $\int x^2 \tan^{-1} x\, dx = \dfrac{x^3}{3}\tan^{-1} x - \dfrac{1}{3}\int \dfrac{x^3}{1+x^2}\, dx$. But $-\dfrac{1}{3}\int \dfrac{x^3}{1+x^2}\, dx = -\dfrac{1}{3}\int\left(x - \dfrac{x}{1+x^2}\right) dx = -\dfrac{1}{3}\left[\dfrac{x^2}{2} - \dfrac{1}{2}\ln\left(1+x^2\right)\right]$. Collecting all related terms, we get the answer as $\int x^2 \tan^{-1} x\, dx = \dfrac{1}{6}\left[2x^3 \tan^{-1} x - x^2 + \ln\left(1+x^2\right)\right]$.

INTEGRAL 34

Problem:

$$\int \frac{2x+3}{\sqrt{3+6x-9x^2}}\,dx$$

Solution:

$$-\frac{1}{9}\left[2\sqrt{3+6x-9x^2}+11\sin^{-1}\left(\frac{1-3x}{2}\right)\right]+constant$$

Techniques used:

Change of variables, Trigonometric identities

Step-by-step solution:

The expression under the radical could be written in the form of an expression containing the term $\left(1-u^2\right)$. Knowing that $\frac{d}{du}\sin^{-1}u=\frac{1}{\sqrt{1-u^2}}$, we can write $3+6x-9x^2=4-\left(1-6x+9x^2\right)=4-\left(1-3x\right)^2=4\left[1-\left(\frac{1-3x}{2}\right)^2\right]$. Therefore, let $\frac{1-3x}{2}=u\Rightarrow dx=-\frac{2}{3}du$. After writing the integral in terms of the variable u, we get $\int\frac{2x+3}{\sqrt{3+6x-9x^2}}\,dx=-\frac{2}{3}\int\frac{2\left(\frac{1-2u}{3}\right)+3}{2\sqrt{1-u^2}}\,du$

$=-\frac{1}{9}\int\frac{11-4u}{\sqrt{1-u^2}}\,du.$ Further, this integral can be written as

$-\frac{1}{9}\int\frac{11-4u}{\sqrt{1-u^2}}\,du=-\frac{11}{9}\int\frac{1}{\sqrt{1-u^2}}\,du+\frac{4}{9}\int\frac{u}{\sqrt{1-u^2}}\,du.$ But $-\frac{11}{9}$

$$\int \frac{1}{\sqrt{1-u^2}} du = -\frac{11}{9} \sin^{-1} u \quad \text{and} \quad \frac{4}{9} \int \frac{u}{\sqrt{1-u^2}} du = -\frac{4}{9}\sqrt{1-u^2}.$$

Collecting all related terms and rewriting the result in terms of the original variable x, we get $-\frac{11}{9} \sin^{-1} u + -\frac{4}{9}\sqrt{1-u^2} =$

$-\frac{11}{9} \sin^{-1}\left(\frac{1-3x}{2}\right) - \frac{4}{9}\sqrt{1-\left(\frac{1-3x}{2}\right)^2}$. After simplification, we get the

answer as $\int \frac{2x+3}{\sqrt{3+6x-9x^2}} dx = -\frac{11}{0} \sin^{-1}\left(\frac{1-3x}{2}\right) - \frac{2}{9}\sqrt{3+6x-9x^2}.$

INTEGRAL 35

Problem:

$$\int \frac{3x+2}{\left(x^2+4\right)\sqrt{x^2+4}}\,dx$$

Solution:

$$\frac{x-6}{2\sqrt{x^2+4}} + constant$$

Techniques used:

Change of variables, Trigonometric identities

Step-by-step solution:

We rewrite the integral as $\int \dfrac{3x+2}{\left(x^2+4\right)\sqrt{x^2+4}}\,dx = 3\int \dfrac{x}{\left(x^2+4\right)^{3/2}}\,dx +$

$2\int \dfrac{1}{\left(x^2+4\right)^{3/2}}\,dx.$ But $3\int \dfrac{x}{\left(x^2+4\right)^{3/2}}\,dx = -3\,/\,\sqrt{\left(x^2+4\right)}.$

To calculate the latter integral, we get a hint from the relation

$\dfrac{d}{dx}\tan u = \dfrac{1}{1+u^2}$ and rewrite $x^2+4 = 4\left[1+\left(\dfrac{x}{2}\right)^2\right]$. Now let $\dfrac{x}{2} =$

$\tan u \Rightarrow dx = \dfrac{2du}{\cos^2 u}.$ Therefore,   we   have $2\int \dfrac{1}{\left(x^2+4\right)^{3/2}}\,dx =$

$2\int \dfrac{2du}{\cos^2 u\left[4\left(1+\tan^2 u\right)\right]^{3/2}} = \dfrac{1}{2}\int \dfrac{du}{\cos^2 u\left(1+\tan^2 u\right)^{3/2}}.$

The denominator of the integrand can be simplified as

$$\cos^2 u \left(1 + \tan^2 u\right)^{3/2} = \cos^2 u \left(\frac{\cos^2 u + \sin^2 u}{\cos^2 u}\right)^{3/2} = \frac{1}{\cos u}. \text{ Therefore,}$$

we have $\displaystyle\frac{1}{2}\int \frac{du}{\cos^2 u \left(1 + \tan^2 u\right)^{3/2}} = \frac{1}{2}\int \cos u\, du = \frac{1}{2}\sin u.$ But

$\sin u = \displaystyle\frac{x}{\sqrt{x^2 + 4}}.$ Collecting all related terms, we get

$$\int \frac{3x + 2}{\left(x^2 + 4\right)\sqrt{x^2 + 4}}\, dx = -\frac{3}{\sqrt{\left(x^2 + 4\right)}} + \frac{1}{2}\frac{x}{\sqrt{x^2 + 4}} = \frac{x - 6}{2\sqrt{\left(x^2 + 4\right)}}.$$

INTEGRAL 36

Problem:

$$\int \frac{2x^2 - 5x - 1}{x^3 - 2x^2 - x + 2} dx$$

Solution:

$$\ln \frac{(x+1)(x-1)^2}{(x-2)} + constant$$

Techniques used:

Partial fractions, Integration by parts

Step-by-step solution:

Using the partial fractions method, the integrand can be written as $\dfrac{2x^2 - 5x - 1}{x^3 - 2x^2 - x + 2} = \dfrac{2x^2 - 5x - 1}{(x-2)(x+1)(x-1)} = \dfrac{A}{x-2} + \dfrac{B}{x+1} + \dfrac{C}{x-1}$.

Therefore, we get $A = -1$, $B = 1$, and $C = 2$. Now rewrite the integral as $\int \dfrac{2x^2 - 5x - 1}{x^3 - 2x^2 - x + 2} dx = -\int \dfrac{1}{x-2} dx + \int \dfrac{1}{x+1} dx + 2\int \dfrac{1}{x-1} dx =$

$-\ln(x-2) + \ln(x+1) + 2\ln(x-1)$. Using the logarithm summation rule, we get the answer as $\ln \dfrac{(x+1)(x-1)^2}{(x-2)}$.

INTEGRAL 37

Problem:

$$\int \frac{2x^3 + 3x^2 + 4}{(x+1)^4} \, dx$$

Solution:

$$2\ln(x+1) + \frac{9x^2 + 18x + 4}{3(x+1)^3} + constant$$

Techniques used:

Partial fractions, Integration by parts, Change of variables

Step-by-step solution:

Let $x + 1 = z \Rightarrow dx = dz$. Therefore, the integral can be written in terms of the variable z as $\int \frac{2x^3 + 3x^2 + 4}{(x+1)^4} \, dx = \int \frac{2(z-1)^3 + 3(z-1)^2 + 4}{z^4} \, dz$.

After expanding the terms in the nominator of the integrand, we get $\int \frac{2(z-1)^3 + 3(z-1)^2 + 4}{z^4} \, dz = \int \frac{2z^3 - 3z^2 + 5}{z^4} \, dz$.

Using the partial fractions technique, we can write the integral as $\int \frac{2z^3 - 3z^2 + 5}{z^4} \, dz = 2\int \frac{dz}{z} - 3\int \frac{dz}{z^2} + 5\int \frac{dz}{z^4} = 2\ln z + \frac{3}{z} - \frac{5}{3z^3}$.

Rewriting the results in terms of the original variable x, we get

$2\ln z + \frac{3}{z} - \frac{5}{3z^3} = 2\ln(x+1) + \frac{3}{x+1} - \frac{5}{3(x+1)^3} = 2\ln(x+1) +$

$\frac{9x^2 + 18x + 4}{3(x+1)^3}$.

INTEGRAL 38

Problem:

$$\int \frac{x-1}{(x+1)\sqrt{x^3+x^2+x}}\,dx$$

Solution:

$$-2\tan^{-1}\left(\frac{\sqrt{x}}{\sqrt{x^2+x+1}}\right) + constant$$

Techniques used:

Change of Variables, Trigonometric identities

Step-by-step solution:

Let $x=z^{-2}\Rightarrow dx=-2z^{-3}dz$. Writing the integral in terms of the variable

z gives, $\displaystyle\int\frac{x-1}{(x+1)\sqrt{x^3+x^2+x}}\,dx=-2\int\frac{\left(z^{-2}-1\right)z^{-3}}{\left(z^{-2}+1\right)\sqrt{z^{-6}+z^{-4}+z^{-2}}}\,dz.$

After multiplying the integrand by $\dfrac{z^5}{z^5}$ we get

$$-2\int\frac{z^5\left(z^{-2}-1\right)z^{-3}}{\underbrace{z^2\left(z^{-2}+1\right)\underbrace{z^3}\sqrt{z^{-6}+z^{-4}+z^{-2}}}_{\sqrt{z^6}}}\,dz=2\int\frac{z^2-1}{\left(z^2+1\right)\sqrt{z^4+z^2+1}}\,dz.$$

Now we let $\tan u=\dfrac{z}{\sqrt{z^4+z^2+1}}$ and differentiate both sides

to get $\dfrac{1}{\cos^2 u}\,du=\dfrac{\sqrt{z^4+z^2+1}-z\left(\dfrac{4z^3+2z}{2\sqrt{z^4+z^2+1}}\right)}{z^4+z^2+1}\,dz.$ Having

$\dfrac{1}{\cos^2 u}=\dfrac{z^4+2z^2+1}{z^4+z^2+1}$, we can write $dz=\dfrac{\left(z^4+2z^2+1\right)\sqrt{z^4+z^2+1}}{1-z^4}\,du.$

After substituting for dz, we can write the integral as

$$2\int \frac{z^2-1}{\left(z^2+1\right)\sqrt{z^4+z^2+1}}\,dz = 2\int\left[\frac{z^2-1}{\left(z^2+1\right)\sqrt{z^4+z^2+1}}\right]$$

$$\underbrace{\left[\frac{\left(z^4+2z^2+1\right)\sqrt{z^4+z^2+1}}{1-z^4}\right]}_{-1}\,du = -2\int du = -2u.$$

But $u = \tan^{-1}\left(\dfrac{z}{\sqrt{z^4+z^2+1}}\right)$ and $z = \dfrac{1}{\sqrt{x}}$. Therefore, we get the answer in terms of the original variable x as $-2u = -2\tan^{-1}$

$$\left(\frac{z}{\sqrt{z^4+z^2+1}}\right) = -2\tan^{-1}\left(\frac{\dfrac{1}{\sqrt{x}}}{\sqrt{\dfrac{1}{x^2}+\dfrac{1}{x}+1}}\right) = -2\tan^{-1}\left(\frac{\sqrt{x}}{\sqrt{x^2+x+1}}\right).$$

INTEGRAL 39

Problem:

$$\int \frac{x^4 + 4x^3 + 6x^2 + 4x + 1}{x^3 - 3x^2 + 3x - 1} dx$$

Solution:

$$24\ln(x-1) + \frac{x^2 + 14x}{2} - \frac{8(4x-3)}{(x-1)^2} + constant$$

Techniques used:

Partial fractions, Change of variables, Integration by parts.

Step-by-step solution:

Using the partial fractions technique, we can write the integral as $\int \frac{x^4 + 4x^3 + 6x^2 + 4x + 1}{x^3 - 3x^2 + 3x - 1} dx = \int (x+7) dx + \int \frac{24x^2 - 16x + 8}{x^3 - 3x^2 + 3x - 1} dx.$

Therefore, we get $\int (x+7) dx = \frac{1}{2}x^2 + 7x$ for the former integral.

For the latter integral, we rewrite this integral as $\int \frac{24x^2 - 16x + 8}{x^3 - 3x^2 + 3x - 1}$

$dx = 8\int \frac{3x^2 - 6x + 3}{x^3 - 3x^2 + 3x - 1} dx = 8\int \frac{3x^2 - 2x + 1 + \overbrace{(4x - 4x)}^{0} + \overbrace{(2 - 2)}^{0}}{x^3 - 3x^2 + 3x - 1} dx =$

$8\int \frac{3x^2 - 6x + 3}{x^3 - 3x^2 + 3x - 1} dx + 16\int \frac{2x - 1}{x^3 - 3x^2 + 3x - 1} dx.$ But

$8\int \frac{3x^2 - 6x + 3}{x^3 - 3x^2 + 3x - 1} dx = 8\ln \underbrace{\left(x^3 - 3x^2 + 3x - 1\right)}_{(x-1)^3} = 24\ln(x-1).$ For

the other integral, we have $16\int \frac{2x - 1}{x^3 - 3x^2 + 3x - 1} dx = 16\int \frac{2x - 1}{(x-1)^3} dx$

we let $x - 1 = z \Rightarrow dx = dz$ and rewrite the integral in terms of the variable z as $16 \int \dfrac{2x - 1}{(x - 1)^3} \, dx = 16 \int \dfrac{2z + 1}{z^3} \, dz$. Using the partial frac-

tion technique, we get $16 \int \dfrac{2z + 1}{z^3} \, dz = 32 \int \dfrac{dz}{z^2} + 16 \int \dfrac{dz}{z^3} = -\dfrac{32}{z} - \dfrac{8}{z^2}$.

Rewriting the obtained results in terms of the original variable x,

we get the answer as $-\dfrac{32}{x - 1} - \dfrac{8}{(x - 1)^2} + 24 \ln(x - 1) + \dfrac{1}{2} x^2 + 7x$

INTEGRAL 40

Problem:

$$\int \frac{1}{\sqrt{x}\left(\sqrt[4]{x}+1\right)^{10}}\,dx$$

Solution:

$$-\frac{1+9\sqrt[4]{x}}{18\left(\sqrt[4]{x}+1\right)^{9}}+constant$$

Techniques used:

Partial fractions, Change of variables

Step-by-step solution:

Let $\sqrt{x}=z \Rightarrow dx=2z\,dz$ and the integral in terms of the variable z, after simplifications, reads $\int \frac{1}{\sqrt{x}\left(\sqrt[4]{x}+1\right)^{10}}\,dx = 2\int \frac{1}{\left(\sqrt{z}+1\right)^{10}}\,dz$. Now, let $\sqrt{z}+1=u \Rightarrow dz=2(u-1)\,du$ and the integral in terms of variable u can be written as $2\int \frac{1}{\left(\sqrt{z}+1\right)^{10}}\,dz = 4\int \frac{u-1}{u^{10}}\,du = 4\int u^{-9}\,du - 4\int u^{-10}\,du$. Performing the integrations gives $-\frac{1}{2}u^{-8}+\frac{4}{9}u^{-9}$. Therefore, the final solution in terms of the original variable x is $-\frac{1}{2}\left(\sqrt{z}+1\right)^{-8}+\frac{4}{9}\left(\sqrt{z}+1\right)^{-9} = -\frac{1}{2}\left(\sqrt[4]{x}+1\right)^{-8}+\frac{4}{9}\left(\sqrt[4]{x}+1\right)^{-9}$. Or, after simplifications, we get the answer as $-\frac{1+9\sqrt[4]{x}}{18\left(\sqrt[4]{x}+1\right)^{9}}$.

INTEGRAL 41

Problem:

$$\int (1+\ln x)\ln(\ln x)dx$$

Solution:

$$x\big[\ln x\ln(\ln x)-1\big]+constant$$

Techniques used:

Change of variables, Integration by parts

Step-by-step solution:

Let $\ln x = z \Rightarrow dx = e^z dz$. Therefore, the integral can be written in terms of the variable z as $\int(1+\ln x)\ln(\ln x)dx = \int(1+z)(\ln z)e^z dz = \int \ln z\, e^z dz + \int(z\ln z)e^z dz$.

But, using the integration by parts technique, we have $\int \ln z\, e^z dz = (z\ln z - z)e^z - \int(z\ln z - z)e^z dz = z\ln z\, e^z - ze^z - \int(z\ln z)e^z dz + \int ze^z dz$. After rearranging the terms, $z\ln z\, e^z - ze^z - \underline{\int(z\ln z)e^z dz} + \int ze^z dz + \int\underline{(z\ln z)e^z dz} = z\ln z\, e^z - ze^z + \int ze^z dz$. The new integral can be worked out using the integration by parts technique as $\int ze^z dz = ze^z - e^z$. Therefore, the answer reads as $z\ln z\, e^z - ze^z + \int ze^z dz = z\ln z\, e^z - ze^z + ze^z - e^z = e^z(z\ln z - 1)$.

Rewriting the solution in terms of the original variable x, we get $e^z(z\ln z - 1) = x\big[\ln x\ln(\ln x)-1\big]$.

INTEGRAL 42

Problem:

$$\int \ln\left(x^2 + x + 1\right) dx$$

Solution:

$$\left(x + \frac{1}{2}\right) \ln\left(x^2 + x + 1\right) - 2x + \sqrt{3} \tan^{-1}\left(\frac{2x+1}{\sqrt{3}}\right) + constant$$

Techniques used:

Change of variables, Integration by parts, Partial fractions

Step-by-step solution:

Using the integration by parts technique, we can write $\int \underbrace{\ln\left(x^2 + x + 1\right)}_{f} \underbrace{dx}_{dg} = x \ln\left(x^2 + x + 1\right) - \int \frac{x(2x+1)}{x^2 + x + 1} dx$. But $\frac{x(2x+1)}{x^2 + x + 1}$

$= \frac{2x^2 + x}{x^2 + x + 1} = 2 - \frac{x+2}{x^2 + x + 1}$, by using the partial fractions technique.

Therefore, $-\int \frac{x(2x+1)}{x^2 + x + 1} dx = -2x + \int \frac{x+2}{x^2 + x + 1} dx$. Now we rewrite

this integral as $\int \frac{x+2}{x^2 + x + 1} dx = \frac{1}{2}\int \frac{2x+1}{x^2 + x + 1} dx + \frac{1}{2}\int \frac{3}{x^2 + x + 1} dx$. But

$\frac{1}{2}\int \frac{2x+1}{x^2 + x + 1} dx = \frac{1}{2}\ln\left(x^2 + x + 1\right)$. Now, for performing the integral

$\frac{1}{2}\int \frac{3}{x^2 + x + 1} dx$, we rewrite the denominator as $x^2 + x + 1 = \left(x + \frac{1}{2}\right)^2$

$+ \frac{3}{4} = \frac{3}{4}\left[1 + \frac{4}{3}\left(x + \frac{1}{2}\right)^2\right] = \frac{3}{4}\left[1 + \left(\frac{2}{\sqrt{3}}x + \frac{1}{\sqrt{3}}\right)^2\right]$. Now, let $\frac{2x+1}{\sqrt{3}} = $

$z \Rightarrow dx = \frac{\sqrt{3}}{2} dz$. Therefore, the integral an be written in terms of the z variable as $\dfrac{1}{2}\displaystyle\int \dfrac{3}{x^2+x+1}\,dx = \dfrac{1}{2}\displaystyle\int \dfrac{3\sqrt{3}}{2\left(\dfrac{3}{4}\right)\left(1+z^2\right)}\,dz =$

$\sqrt{3}\displaystyle\int \dfrac{dz}{1+z^2} = \sqrt{3}\tan^{-1} z = \sqrt{3}\tan^{-1}\left(\dfrac{2x+1}{\sqrt{3}}\right)$. Collecting all the

calculated related terms, we get the answer as $x\ln\left(x^2+x+1\right)-$

$2x+\dfrac{1}{2}\ln\left(x^2+x+1\right)+\sqrt{3}\tan^{-1}\left(\dfrac{2x+1}{\sqrt{3}}\right).$

INTEGRAL 43

Problem:

$$\int \cos x \sqrt{4 - \sin^2 x} \, dx$$

Solution:

$$2 \left[\sin^{-1} \left(\frac{\sin x}{2} \right) + \frac{\sin x}{4} \sqrt{4 - \sin^2 x} \right] + constant$$

Techniques used:

Change of variables, Integration by parts, Trigonometric identities

Step-by-step solution:

Let $\sin x = 2 \sin z \Rightarrow \cos x \, dx = 2 \cos z \, dz$. We can rewrite the integral in terms of the variable z as $\int \cos x \sqrt{4 - \sin^2 x} \, dx = 4 \int \cos z \sqrt{1 - \sin^2 z} \, dz$

$= 4 \int \cos^2 z \, dz$. But $\cos^2 z = \frac{1 + \cos 2z}{2}$. Therefore, the integral can be

written as $4 \int \cos^2 z \, dz = 2 \int (1 + \cos 2z) \, dz = 2z + \sin 2z$. Now writing

this result in terms of the original variable x, we get $2z + \sin 2z =$

$2 \sin^{-1} \left(\frac{\sin x}{2} \right) + \sin x \sqrt{1 - \left(\frac{\sin x}{2} \right)^2}$. Note that $z = \sin^{-1} \left(\frac{\sin x}{2} \right)$ and

$\sin 2z = 2 \sin z \cos z$.

INTEGRAL 44

Problem:

$$\int \frac{\cos x}{\sin^2 x - 3\sin x + 2} dx$$

Solution:

$$\ln\left(\frac{\sin x - 2}{\sin x - 1}\right) + constant$$

Techniques used:

Change of variables, Integration by parts, Partial fractions

Step-by-step solution:

Let $\sin x = z \Rightarrow \cos x\,dx = dz$ and rewrite the integral in terms of the variable z to get $\int \dfrac{\cos x}{\sin^2 x - 3\sin x + 2} dx = \int \dfrac{1}{z^2 - 3z + 2} dz$. But $z^2 - 3z + 2 = (z-2)(z-1)$. Therefore, the integral can be written as, using the partial fractions technique, $\int \dfrac{1}{z^2 - 3z + 2} dz =$

$\int \dfrac{1}{(z-2)(z-1)} dz = \int \dfrac{1}{z-2} dz - \int \dfrac{1}{z-1} dz.$ Performing the integration

of the integrals gives $\int \dfrac{1}{z-2} dz - \int \dfrac{1}{z-1} dz = \ln(z-2) - \ln(z-1) =$

$\ln \dfrac{z-2}{z-1}$. After writing the results in terms of the original variable x

we get $\ln\left(\dfrac{\sin x - 2}{\sin x - 1}\right)$.

INTEGRAL 45

Problem:

$$\int \frac{1}{\left(1+\sqrt{x}\right)\sqrt{x-x^2}} dx$$

Solution:

$$2\frac{\sqrt{1-\sqrt{x}}}{\sqrt{1+\sqrt{x}}} + constant$$

Techniques used:

Change of variables

Step-by-step solution:

Let $\sqrt{x} = z \Rightarrow dx = 2zdz$. Rewriting the integral in terms of the variable z we get, after some simplifications, $\int \frac{1}{\left(1+\sqrt{x}\right)\sqrt{x-x^2}} dx =$

$2\int \frac{1}{\left(1+z\right)\sqrt{1-z^2}} dz$. Now, let $\frac{\sqrt{1-z}}{\sqrt{1+z}} = u \Rightarrow dz = \frac{-\left(1+z\right)^2\sqrt{1-z}}{\sqrt{1+z}} du$.

Substituting into the integral (writing the integrand in terms of both z and u variables), we have $2\int \frac{1}{\left(1+z\right)\sqrt{1-z^2}} dz =$

$-2\int \underbrace{\frac{1}{\left(1+z\right)\sqrt{1-z^2}} \frac{\left(1+z\right)^2\sqrt{1-z}}{\sqrt{1+z}}}_{=1} du = -2u$. Therefore, the answer

in terms of the original variable x can be written as

$-2u = -2\frac{\sqrt{1-z}}{\sqrt{1+z}} = -2\frac{\sqrt{1-\sqrt{x}}}{\sqrt{1+\sqrt{x}}}$.

INTEGRAL 46

Problem:

$$\int \frac{1 + 2x^2}{x^5 \left(1 + x^2\right)^3} \, dx$$

Solution:

$$\frac{-1}{4x^4 \left(1 + x^2\right)^2} + constant$$

Techniques used:

Change of variables, Integration by parts

Step-by-step solution:

The denominator of the integrand can be written as $x^5 \left(1 + x^2\right)^3 =$

$\frac{1}{x}\left(x^6\right)\left(1 + x^2\right)^3 = \frac{1}{x}\left(x^2 + x^4\right)^3$. Noticing that $\frac{d}{dx}\left(x^2 + x^4\right) = 2x + 4x^3$, a multiple of the numerator after multiplying it by x. Now let $x^2 + x^4 = z \Rightarrow dx = \frac{dz}{2\left(x + 2x^3\right)}$. Therefore, the integral can be written in terms of the variable z as $\int \frac{1 + 2x^2}{x^5 \left(1 + x^2\right)^3} \, dx = \int \frac{x + 2x^3}{\left(x^2 + x^4\right)^3} \frac{dz}{2\left(x + 2x^3\right)} =$

$\frac{1}{2}\int \frac{1}{z^3} \, dz = \frac{-1}{4z^2}$. Or, in terms of the original variable x we get

$\frac{-1}{4z^2} = \frac{-1}{4\left(x^2 + x^4\right)^2} = \frac{-1}{4x^4 \left(1 + x^2\right)^2}$.

Please note that it is possible to calculate this integral without using the tip mentioned above. But it will be a lengthier operation.

INTEGRAL 47

Problem:

$$\int \sin^6 x \cos^5 x dx$$

Solution:

$$\frac{\sin^7 x}{693}\left(63\sin^4 x - 154\sin^2 x + 99\right) + constant$$

Techniques used:

Change of variables, Trigonometric identities

Step-by-step solution:

The integrand can be written, using trigonometric identities, as
$\sin^6 x \cos^5 x = \sin^6 x \underbrace{\left(1 - \sin^2 x\right)^2}_{\cos^4 x} \cos x$. Now let $\sin x = z \Rightarrow \cos x dx = dz$. Therefore, the integral can be written in terms of variable z as
$\int \sin^6 x \cos^5 x \, dx = \int \sin^6 x \left(1 - \sin^2 x\right)^2 \cos x \, dx = \int z^6 \left(1 - z^2\right)^2 dz$. After
expanding the integral we get $\int z^6 \left(1 - z^2\right)^2 dz = \int z^{10}dz - 2\int z^8 dz + \int z^6 dz$. Performing the integration operation for each term gives,
$\frac{1}{11}z^{11} - \frac{2}{9}z^9 + \frac{1}{7}z^7$. Rewriting the results in terms of the original
variable x gives $\frac{1}{11}\sin^{11} x - \frac{2}{9}\sin^9 x + \frac{1}{7}\sin^7 x$. Or simplify to have
$\frac{\sin^7 x}{693}\left(63\sin^4 x - 154\sin^2 x + 99\right)$.

Note that this integral is a version of the general recursive formula, given as:

$$\int \sin^m ax \cos^n ax \, dx = \frac{\sin^{m+1} ax \cos^{n-1} ax}{a(m+n)} + \frac{n-1}{m+n}\int \sin^m ax \cos^{n-2} ax \, dx$$

We calculated this integral as given by Integral 24, using only integration by parts technique. Readers can verify that the result given in this section and that of section 24 are equivalent.

INTEGRAL 48

Problem:

$$\int x^2 e^{x^{3/2}}\, dx$$

Solution:

$$\frac{2}{3} e^{x^{3/2}} \left(x^{3/2} - 1 \right) + constant$$

Techniques used:

Change of variables, Integration by parts

Step-by-step solution:

Let $x^{3/2} = z \Rightarrow dx = \dfrac{2}{3\sqrt{x}} dz$. Therefore, the integral can be written

as $\int x^2 e^{x^{3/2}} dx = \dfrac{2}{3}\int \underbrace{\left(\dfrac{x^2}{\sqrt{x}} \right)}_{x^{1.5}=z} e^z dz = \dfrac{2}{3}\int z e^z dz$. Using the integration by

parts technique we get $\dfrac{2}{3}\int z e^z dz = \dfrac{2}{3}\left(z e^z - e^z \right)$. Or in terms of the

original variable x, we have $\dfrac{2}{3} e^{x^{3/2}} \left(x^{3/2} - 1 \right)$.

INTEGRAL 49

Problem:

$$\int \frac{\tan^3 x}{\cos^3 x} dx$$

Solution:

$$\frac{3 - 5\cos^2 x}{15\cos^5 x} + constant$$

Techniques used:

Change of variables, Integration by parts

Step-by-step solution:

Let $\dfrac{1}{\cos x} = z \Rightarrow dx = \dfrac{dz}{z^2 \sin x}$. Therefore, the integral can be written in terms of the variable z as $\displaystyle\int \frac{\tan^3 x}{\cos^3 x} dx = \int \frac{\sin^3 x}{\cos^3 x} z^3 \left(\frac{dz}{z^2 \sin x}\right) =$

$\displaystyle\int z^2 (z^2 - 1) dz$. Performing the integration operation for the

latter integral gives $\displaystyle\int z^2 (z^2 - 1) dz = \int z^4 dz - \int z^2 dz = \frac{z^5}{5} - \frac{z^3}{3}$.

Writing the results in terms of the original variable x, we get

$\dfrac{z^5}{5} - \dfrac{z^3}{3} = \dfrac{1}{5\cos^5 x} - \dfrac{1}{3\cos^3 x} = \dfrac{1}{15\cos^5 x}(3 - 5\cos^2 x) = \dfrac{1}{15\cos^5 x}$

$\left(3\sin^2 x - 2\cos^2 x\right)$. This expression simplifies to $\dfrac{3 - 5\cos^2 x}{15\cos^5 x}$.

INTEGRAL 50

Problem:

$$\int \frac{\sqrt[3]{\tan x}}{\left(\sin x + \cos x\right)^2}\,dx$$

Solution:

$$\frac{\sqrt{3}}{3}\tan^{-1}\left(\frac{2\sqrt[3]{\tan x}-1}{\sqrt{3}}\right) - \frac{\sqrt[3]{\tan x}}{1+\tan x} - \frac{1}{6}\ln\left(\sqrt[3]{\tan^2 x} - \sqrt[3]{\tan x} + 1\right)$$

$$+ \frac{1}{3}\ln\left(1 + \sqrt[3]{\tan x}\right) + constant$$

Techniques used:

Change of variables, Integration by parts

Step-by-step solution:

Let $\tan x = z \Rightarrow dx = \cos^2 x\,dz$, and write the integral in terms of variable z to get $\int \frac{\sqrt[3]{\tan x}}{\left(\sin x + \cos x\right)^2}\,dx = \int \frac{\sqrt[3]{\tan x}}{\left(1+\tan x\right)^2 \cos^2 x}\,dx =$

$\int \frac{\sqrt[3]{z}}{\left(1+z\right)^2}\,dz$. Now let $z = u^3 \Rightarrow dz = 3u^2 du$, and write the integral in terms of variable u along with applying the integration by parts technique, to get $\int \frac{\sqrt[3]{z}}{\left(1+z\right)^2}\,dz = \int \underbrace{\frac{3u^2}{\left(1+u^3\right)^2}}_{dg}\,\overset{u}{\overbrace{u}}\,du = \frac{-u}{1+u^3} + \int \frac{du}{1+u^3}.$

Hence, we can write $\frac{-u}{1+u^3} = \frac{-\sqrt[3]{z}}{1+z} = \frac{-\sqrt[3]{\tan x}}{1+\tan x}$. For the latter integral we use the partial fractions technique to have $\int \frac{du}{1+u^3} = \frac{1}{3}\int \frac{1}{1+u}\,du + \frac{1}{3}\int \frac{2-u}{u^2-u+1}\,du$. The former integral reads

$\frac{1}{3}\int\frac{du}{1+u} = \frac{1}{3}\ln(1+u)$. Or $\frac{1}{3}\ln(1+u) = \frac{1}{3}\ln\left(1+\sqrt[3]{\tan x}\right)$. Rewrite

the latter integral as $\frac{1}{3}\int\frac{2-u}{u^2-u+1}du = \frac{1}{6}\int\frac{1-2u}{u^2-u+1}du +$

$\frac{1}{2}\int\frac{1}{u^2-u+1}du$. But we have $\frac{1}{6}\int\frac{1-2u}{u^2-u+1}du = -\frac{1}{6}\ln(u^2-u+1)$

$= -\frac{1}{6}\ln\left(\sqrt[3]{z^2}-\sqrt[3]{z}+1\right) = \frac{-1}{6}\ln\left(\sqrt[3]{\tan^2 x}-\sqrt[3]{\tan x}+1\right).$ For

the remaining integral, we rewrite it as $\frac{1}{2}\int\frac{1}{u^2-u+1}du =$

$$\frac{1}{2}\int\frac{1}{(u-1/2)^2+3/4}du = \frac{1}{2}\cdot\frac{4}{3}\int\frac{1}{1+\left(\frac{2u-1}{\sqrt3}\right)^2}du = \frac{2}{3}\int\frac{1}{1+\left(\frac{2u-1}{\sqrt3}\right)^2}du.$$

Now let $\frac{2u-1}{\sqrt3} = y \Rightarrow du = \frac{\sqrt3}{2}dy$, and write the integral in terms

of variable y. Hence $\frac{2}{3}\int\frac{1}{1+\left(\frac{2u-1}{\sqrt3}\right)^2}du = \frac{\sqrt3}{3}\int\frac{1}{1+y^2}dy$. But we

have $\frac{\sqrt3}{3}\int\frac{1}{1+y^2} = \frac{\sqrt3}{3}\tan^{-1}y$. Or, $\frac{\sqrt3}{3}\tan^{-1}y = \frac{\sqrt3}{3}\tan^{-1}\left(\frac{2u-1}{\sqrt3}\right) =$

$\frac{\sqrt3}{3}\tan^{-1}\left(\frac{2\sqrt[3]{z}-1}{\sqrt3}\right) = \frac{\sqrt3}{3}\tan^{-1}\left(\frac{2\sqrt[3]{\tan x}-1}{\sqrt3}\right).$ Collecting all

answers, we get the solutions as $-\frac{\sqrt[3]{\tan x}}{1+\tan x} + \frac{1}{3}\ln\left(1+\sqrt[3]{\tan x}\right) -$

$\frac{1}{6}\ln\left(\sqrt[3]{\tan^2 x}-\sqrt[3]{\tan x}+1\right) + \frac{\sqrt3}{3}\tan^{-1}\left(\frac{2\sqrt[3]{\tan x}-1}{\sqrt3}\right).$

INTEGRAL 51

Problem:

$$\int \left(\frac{1}{\ln x} - \frac{1}{\ln^2 x} \right) dx$$

Solution:

$$\frac{x}{\ln x} + constant$$

Techniques used:

Integration by parts

Step-by-step solution:

Rewriting the integral as sum of two parts, we get $\int \left(\dfrac{1}{\ln x} - \dfrac{1}{\ln^2 x} \right) dx =$
$\int \dfrac{dx}{\ln x} - \int \dfrac{dx}{\ln^2 x}$. But the latter integral can be calculated, using the integration by parts as $\int \dfrac{dx}{\ln^2 x} = \int \dfrac{xdx}{x\ln^2 x} = \int \underset{f}{x} \underset{dg}{\underbrace{\left(\dfrac{dx}{x\ln^2 x} \right)}} = -\dfrac{x}{\ln x} + \int \dfrac{1}{\ln x} dx$.

Note that $\dfrac{d}{dx}\left(\dfrac{1}{\ln x} \right) = -\dfrac{1}{x\ln^2 x}$. Now, by collecting all terms we have
$\int \dfrac{dx}{\ln x} - \int \dfrac{dx}{\ln^2 x} = \int \dfrac{1}{\ln x} dx + \dfrac{x}{\ln x} - \int \dfrac{1}{\ln x} dx = \dfrac{x}{\ln x}$.

INTEGRAL 52

Problem:

$$\int \left[\sin(x+\sin x) - \sin(x-\sin x)\right]dx$$

Solution:

$$-2\cos(\sin x) + constant$$

Techniques used:

Change of variables, Trigonometry identities

Step-by-step solution:

Expand the integrand, $\sin(x+\sin x) - \sin(x-\sin x) = \sin x \cos(\sin x)$ $+ \cos x \sin(\sin x) - \sin x \cos(\sin x) + \cos x \sin(\sin x) = 2\cos x$ $\sin(\sin x)$. Let $\sin x = z \Rightarrow \cos x dx = dz$ and rewrite the integral in terms of the variable z, to get $2\int \cos x \sin(\sin x)dx = 2\int \sin z dz = -2\cos z$. Rewriting the results in terms of the original variable x, we get $-2\cos(\sin x)$.

INTEGRAL 53

Problem:

$$\int \frac{x}{1-5x^2} \sqrt{\left(\frac{2}{1+5x^2}-1\right)}\, dx$$

Solution:

$$\frac{1}{10}\sin^{-1}\left(5x^2\right)+constant$$

Techniques used:

Change of variables, Integration by parts

Step-by-step solution:

Manipulating the integrand gives, $\dfrac{x}{1-5x^2}\sqrt{\left(\dfrac{2}{1+5x^2}-1\right)}=$

$\dfrac{x}{1-5x^2}\sqrt{\dfrac{1-5x^2}{1+5x^2}}=\dfrac{x}{\sqrt{1-25x^4}}$. Let $5x^2=z \Rightarrow 10x\,dx=dz$ and write

the integral in terms of the variable z, to get $\displaystyle\int \frac{x}{\sqrt{1-25x^4}}dx=$

$\dfrac{1}{10}\displaystyle\int \frac{1}{\sqrt{1-z^2}}dz$. But $\displaystyle\int \frac{1}{\sqrt{1-z^2}}=\sin^{-1}z$. Therefore, the result in

terms of the original variable x reads $\dfrac{1}{10}\displaystyle\int \frac{1}{\sqrt{1-z^2}}dz=$

$\dfrac{1}{10}\sin^{-1}z=\dfrac{1}{10}\sin^{-1}\left(5x^2\right).$

INTEGRAL 54

Problem:

$$\int \frac{1}{\left(x+\dfrac{1}{x}\right)^2}\,dx$$

Solution:

$$\frac{1}{2}\left(\tan^{-1}x-\frac{x}{1+x^2}\right)+constant$$

Techniques used:

Integration by parts, Change of variables, Trigonometric identities

Step-by-step solution:

Manipulating the integrand gives $\dfrac{1}{\left(x+\dfrac{1}{x}\right)^2}=\dfrac{x^2}{\left(1+x^2\right)^2}=\dfrac{x^2+\overset{0}{\overbrace{1-1}}}{\left(1+x^2\right)^2}=$

$\dfrac{1}{1+x^2}-\dfrac{1}{\left(1+x^2\right)^2}$. Therefore, the integral can be written as

$\displaystyle\int\frac{1}{\left(x+\dfrac{1}{x}\right)^2}\,dx=\int\frac{1}{1+x^2}\,dx-\int\frac{1}{\left(1+x^2\right)^2}\,dx$ But $\displaystyle\int\frac{1}{1+x^2}\,dx=\tan^{-1}x.$

For performing the latter integral, let $x=\tan z\Rightarrow dx=\dfrac{dz}{\cos^2 z}$ and

rewrite this integral in terms of the variable z to get $\displaystyle\int\frac{1}{\left(1+x^2\right)^2}\,dx$

$\displaystyle=\int\frac{1}{\left(1+\tan^2 z\right)^2}\frac{dz}{\cos^2 z}=\int\cos^2 z\,dz.$ But $\cos^2 z=\dfrac{1+\cos(2z)}{2}.$ There-

fore, $\displaystyle\int\cos^2 z\,dz=\frac{1}{2}\int\left(1+\cos(2z)\right)dz=\frac{1}{2}z+\frac{1}{4}\sin 2z.$ Collecting all

related terms and rewriting the results in terms of the

original variable x gives $\tan^{-1} x - \dfrac{1}{2}\tan^{-1} x - \dfrac{1}{4}\sin\left(2\tan^{-1} x\right) =$

$\dfrac{1}{2}\tan^{-1} x - \dfrac{1}{4}\sin\left(2\tan^{-1} x\right)$. But the last term in the latter expres-

sion can be written as $\dfrac{1}{4}\sin\left(\underbrace{2\tan^{-1} x}_{\alpha}\right) = \dfrac{1}{4}(\sin 2\alpha) = \dfrac{1}{2}\sin\alpha\cos\alpha$.

But, having $\tan\alpha = x$, we get $\dfrac{1}{2}\sin\alpha\cos\alpha = \dfrac{x}{2+2x^2}$. Therefore, the

results of the integral can be written as $\dfrac{1}{2}\tan^{-1} x - \dfrac{1}{4}\sin\left(2\tan^{-1} x\right) =$

$\dfrac{1}{2}\left(\tan^{-1} x - \dfrac{x}{1+x^2}\right)$.

INTEGRAL 55

Problem:

$$\int e^{2x} \tan^{-1}\left(e^x\right) dx$$

Solution:

$$\frac{1}{2}\left[\left(1+e^{2x}\right)\tan^{-1}\left(e^x\right) - e^x\right] + constant$$

Techniques used:

Chane of variables, Integration by parts

Step-by-step solution:

Let $e^x = z \Rightarrow dx = dz / z$, and rewrite the integral in terms of the variable z to get $\int e^{2x} \tan^{-1}\left(e^x\right) dx = \int z \tan^{-1} z \, dz$. Using the integration by parts technique, we get $\int \underbrace{z}_{dg} \underbrace{\tan^{-1} z}_{f} \, dz = \frac{z^2}{2}\tan^{-1} z - \frac{1}{2}\int \frac{z^2}{1+z^2} dz$.

Note that $\dfrac{d}{dz}\tan^{-1} z = \dfrac{1}{1+z^2}$. Therefore, $\dfrac{z^2}{2}\tan^{-1} z = \dfrac{1}{2}\left(e^{2x}\tan^{-1} e^x\right)$.
To calculate the remaining integral, rewrite it as
$-\dfrac{1}{2}\int \dfrac{z^2}{1+z^2} dz = -\dfrac{1}{2}\int \dfrac{z^2+1-1}{1+z^2} dz = -\dfrac{1}{2}\int dz + \dfrac{1}{2}\int \dfrac{1}{1+z^2} dz =$
$\dfrac{1}{2}\left(-z + \tan^{-1} z\right) = \dfrac{1}{2}\left(-e^x + \tan^{-1} e^x\right)$. After collecting all obtained results, as underlined, and simplifying we get the solution as
$\dfrac{1}{2}\left[\left(1+e^{2x}\right)\tan^{-1} e^x - e^x\right]$.

INTEGRAL 56

Problem:

$$\int \frac{1}{x\sqrt{6x - x^2}}\, dx$$

Solution:

$$-\frac{\sqrt{6x - x^2}}{3x} + constant$$

Techniques used:

Change of variables, Integration by parts

Step-by-step solution:

The expression $6x - x^2$, can be written as $6x - x^2 = 9 - (x - 3)^2 =$
$9\left[1 - \left(\frac{x}{3} - 1\right)^2\right]$. Substituting into the integral, we get $\int \frac{1}{x\sqrt{6x - x^2}}\, dx =$

$\frac{1}{3}\int \frac{1}{x\sqrt{1 - \left(\frac{x}{3} - 1\right)^2}}\, dx$. Now, let $\frac{x}{3} - 1 = z \Rightarrow dx = 3dz$ and rewrite the

integral in terms of the variable z to get $\frac{1}{3}\int \frac{1}{x\sqrt{1 - \left(\frac{x}{3} - 1\right)^2}}\, dx =$

$\frac{1}{3}\int \frac{1}{(z+1)\sqrt{1 - z^2}}\, dz$. Now, Let and $\sqrt{1 - z^2} = \cos u$. Therefore,

$\frac{1}{3}\int \frac{1}{(z+1)\sqrt{1 - z^2}}\, dz = \frac{1}{3}\int \frac{1}{1 + \sin u}\, du$. Using the trigonometric

identity $\sin u = \frac{2\tan(u/2)}{1 + \tan^2(u/2)}$, we get $\frac{1}{3}\int \frac{1}{1 + \dfrac{2\tan(u/2)}{1 + \tan^2(u/2)}}\, du =$

$$\frac{1}{3}\int \frac{1+\tan^2(u/2)}{\left(1+\tan(u/2)\right)^2}du = \frac{1}{3}\int \frac{1/\cos^2(u/2)}{\left(1+\tan(u/2)\right)^2}du.$$

Let $1+\tan(u/2)=y \Rightarrow du = 2\cos^2(u/2)dy$, and the solution of the integral in terms of the variable y reads, after substitution,

$$\frac{1}{3}\int \frac{1/\cos^2(u/2)}{\left(1+\tan(u/2)\right)^2}du = \frac{2}{3}\int \frac{1/\cos^2(u/2)}{y^2}\cos^2(u/2)dy = \frac{2}{3}\int \frac{1}{y^2}dy$$

$$= -\frac{2}{3y}.$$ The answer can be written in terms of the original variable

u, after substitution, as $-\dfrac{2}{3y} = -\dfrac{2}{3\left(1+\tan\left(\dfrac{u}{2}\right)\right)}.$ But $\tan\left(\dfrac{u}{2}\right)=$

$$\frac{\sin(u/2)}{\cos(u/2)} = \frac{\sin u}{2\cos^2(u/2)} = \frac{\sin u}{1+\cos u},\quad \text{and}\quad -\frac{2}{3\left(1+\tan\left(\dfrac{u}{2}\right)\right)} =$$

$-\dfrac{2(1+\cos u)}{3(1+\sin u+\cos u)}.$ Therefore, in terms of the variable z, having

$z=\sin u$, we get $-\dfrac{2(1+\cos u)}{3(1+\sin u+\cos u)} = -\dfrac{2\left(1+\sqrt{1-z^2}\right)}{3\left(1+z+\sqrt{1-z^2}\right)}.$ Finally,

in terms of the original variable x, we get $-\dfrac{2\left(1+\sqrt{1-z^2}\right)}{3\left(1+z+\sqrt{1-z^2}\right)} =$

$$-\frac{2\left(1+\sqrt{1-\left(\dfrac{x}{3}-1\right)^2}\right)}{3\left(1+\dfrac{x}{3}-1+\sqrt{1-\left(\dfrac{x}{3}-1\right)^2}\right)}.$$ This expression simplifies to

$$-\frac{x+\sqrt{6x-x^2}}{3x} = -\frac{1}{3} - \frac{\sqrt{6x-x^2}}{3x}.$$

INTEGRAL 57

Problem:

$$\int \tan^3 x \, dx$$

Solution:

$$\frac{1}{2} \tan^2 x + \ln(\cos x) + constant$$

Techniques used:

Integration by parts, Trigonometric identities

Step-by-step solution:

Having $\dfrac{d}{dx}(\tan x) = \dfrac{1}{\cos^2 x}$, we use the integration by parts tech-

nique to get $\int \tan^3 x \, dx = \int \underbrace{\left(\dfrac{dx}{\cos^2 x}\right)}_{dg} \underbrace{\dfrac{\sin^3 x}{\cos x}}_{f} = \tan x \left(\dfrac{\sin^3 x}{\cos x}\right) - \int \tan x$

$\left(\dfrac{3\cos^2 x \sin^2 x + \sin^4 x}{\cos^2 x}\right)$. But the integrand, after expanding

and some manipulations, reads $\dfrac{\sin x}{\cos x} \left(\dfrac{3\cos^2 x \sin^2 x + \sin^4 x}{\cos^2 x}\right) =$

$\dfrac{\sin^3 x}{\cos^3 x} \left(3\cos^2 x + \sin^2 x\right) = \tan^3 x \left(3\cos^2 x + 1 - \cos^2 x\right) = \tan^3 x +$

$2 \tan^3 x \cos^2 x$. After back substitution, we get $\int \tan^3 x \, dx =$

$\tan x \left(\dfrac{\sin^3 x}{\cos x}\right) - \int \left(\tan^3 x + 2\tan^3 x \cos^2 x\right) dx = \tan x \left(\dfrac{\sin^3 x}{\cos x}\right) - \int \tan^3$

$x - 2 \int \tan^3 x \cos^2 x$. Or, after rearranging the terms in the last equation, we get $\int \tan^3 x dx = \frac{1}{2} \tan x \left(\frac{\sin^3 x}{\cos x} \right) - \int \tan^3 x \cos^2 x$. But

$-\int \tan^3 x \cos^2 x = -\int \frac{\sin x}{\cos x} \left(\sin^2 x \right) dx = -\int \frac{\sin x}{\cos x} dx + \int \sin x \cos x dx$.

But $-\int \frac{\sin x}{\cos x} dx = \ln(\cos x)$ and $\int \sin x \cos x dx = \frac{1}{2} \sin^2 x$. Therefore, the final answer, after collecting all related terms, reads $\int \tan^3 x dx = \frac{1}{2} \tan x \left(\frac{\sin^3 x}{\cos x} \right) + \ln(\cos x) + \frac{1}{2} \sin^2 x$. The answer can be simplified to $\frac{1}{2} \tan x \left(\frac{\sin^3 x}{\cos x} \right) + \ln(\cos x) + \frac{1}{2} \sin^2 x = \frac{1}{2} \sin^2$

$x \left(1 + \frac{\sin^2 x}{\cos^2 x} \right) + \ln(\cos x) = \frac{1}{2} \tan^2 x + \ln(\cos x)$.

INTEGRAL 58

Problem:

$$\int \frac{1}{4+5\cos x}\,dx$$

Solution:

$$\frac{2}{3}\tanh^{-1}\left[\frac{\tan(x/2)}{3}\right] + constant$$

Techniques used:

Change of variables, Integration by parts, Trigonometric identities

Step-by-step solution:

Let $\tan\left(\dfrac{x}{2}\right) = z \Rightarrow dx = 2\cos^2\left(\dfrac{x}{2}\right)dz$. Now, using the trigonometric

identity $\cos x = \dfrac{1-\tan^2\dfrac{x}{2}}{1+\tan^2\dfrac{x}{2}}$, rewrite the integral in terms of variable z

as $\displaystyle\int \frac{1}{4+5\cos x}\,dx = 2\int \frac{\cos^2\left(\dfrac{x}{2}\right)}{4+5\left(\dfrac{1-\tan^2\dfrac{x}{2}}{1+\tan^2\dfrac{x}{2}}\right)}\,dz = 2\int \frac{1}{9-z^2}\,dz.$

But $\displaystyle 2\int \frac{1}{9-z^2}\,dz = \frac{2}{9}\int \frac{1}{1-\left(\dfrac{z}{3}\right)^2}\,dz.$ Now let $\dfrac{z}{3} = u \Rightarrow dz = 3du$, and

rewriting the latter integral in terms of the variable u gives

$\displaystyle \frac{2}{9}\int \frac{1}{1-\left(\dfrac{z}{3}\right)^2}\,dz = \frac{2}{3}\int \frac{1}{1-u^2}\,du.$ But $\displaystyle\int \frac{1}{1-u^2}\,du = \tanh^{-1} u$. Therefore,

writing the answer in terms of variables z and then x gives

$$\frac{2}{3}\tanh^{-1} u = \frac{2}{3}\tanh^{-1}\left(\frac{z}{3}\right) = \frac{2}{3}\tanh^{-1}\left(\frac{\tan(x/2)}{3}\right).$$

INTEGRAL 59

Problem:

$$\int \frac{\tan^3(1+\ln x)}{x}\,dx$$

Solution:

$$\frac{1}{2}\tan^2(1+\ln x)+\ln\left[\cos(1+\ln x)\right]+constant$$

Techniques used:

Change of variables, Integration by parts, Trigonometric identities

Step-by-step solution:

Let $\ln x = z \Rightarrow dx = e^z dz$ and rewrite the integral in terms of the variable z, to get $\int \frac{\tan^3(1+\ln x)}{x}\,dx = \int \frac{\tan^3(1+z)}{e^z}e^z dz = \int \tan^3(1+z)\,dz.$

Now using the results from Integral 57, we can write the answer as $\frac{1}{2}\tan^2(1+z)+\ln(\cos(1+z))$. Or in terms of the original variable x, we have $\frac{1}{2}\tan^2(1+\ln x)+\ln(\cos(1+\ln x))$.

INTEGRAL 60

Problem:

$$\int \left(\frac{\sin 2x \sin 3x}{\sin x \sin 6x} \right)^2 dx$$

Solution:

$$\frac{\sqrt{3}}{18} \left[\ln \left(\frac{1 + \sqrt{3} \tan x}{1 - \sqrt{3} \tan x} \right) + \frac{1}{1 - 3\tan^2 x} \right] + constant$$

Techniques used:

Integration by parts, Trigonometric identities, Change of variables

Step-by-step solution:

Simplify the expression inside the bracket of the integrand to $\frac{\sin 2x \sin 3x}{\sin x \sin 6x} = \frac{2\sin x \cos x \sin 3x}{2\sin x \sin 3x \cos 3x} = \frac{\cos x}{\cos 3x}$. But we have $\cos 3x = 4\cos^3 x - 3\cos x$. Therefore, the integral can be written as $\int \left(\frac{\sin 2x \sin 3x}{\sin x \sin 6x} \right)^2 dx = \int \frac{1}{\left(4\cos^2 x - 3\right)^2} dx$. Now, let $\tan x = z \Rightarrow dx = \cos^2 x dz = \frac{dz}{1+z^2}$ and write the integral in terms of the variable z as $\int \frac{1}{\left(4\cos^2 x - 3\right)^2} dx = \int \frac{1+z^2}{\left(1-3z^2\right)^2} dz$. Using the partial fractions technique, we get $\int \frac{1+z^2}{\left(1-3z^2\right)^2} dz = \int \frac{1+z^2}{\left(1-\sqrt{3}z\right)^2 \left(1+\sqrt{3}z\right)^2} dz$

$\frac{1}{6} \int \frac{dz}{1-\sqrt{3}z} + \frac{1}{3} \int \frac{dz}{\left(1+\sqrt{3}z\right)^2} + \frac{1}{3} \int \frac{dz}{\left(1-\sqrt{3}z\right)^2}$. The first two integrals in the latter expression read $\dfrac{1}{6}\displaystyle\int\dfrac{dz}{1+\sqrt{3}z}+\dfrac{1}{6}\displaystyle\int\dfrac{dz}{1-\sqrt{3}z}=$

$\dfrac{1}{6\sqrt{3}}\ln\left(1+\sqrt{3}z\right)-\dfrac{1}{6\sqrt{3}}\ln\left(1-\sqrt{3}z\right)=\dfrac{1}{6\sqrt{3}}\ln\left(\dfrac{1+\sqrt{3}z}{1-\sqrt{3}z}\right)=$

$\underline{\dfrac{1}{6\sqrt{3}}\ln\left(\dfrac{1+\sqrt{3}\tan x}{1-\sqrt{3}\tan x}\right)}$. For the remaining integral, let $1+\sqrt{3}z=$

$u\Rightarrow dz=\dfrac{du}{\sqrt{3}}$ and write the integral in terms of variable

u as $\dfrac{1}{3}\displaystyle\int\dfrac{dz}{\left(1+\sqrt{3}z\right)^{2}}=\dfrac{1}{3\sqrt{3}}\displaystyle\int\dfrac{du}{u^{2}}=-\dfrac{1}{3\sqrt{3}u}=-\dfrac{1}{3\sqrt{3}\left(1+\sqrt{3}z\right)}=$

$-\dfrac{1}{3\sqrt{3}\left(1+\sqrt{3}\tan x\right)}$. Or after simplification reads $\underline{\dfrac{-\sqrt{3}}{9\left(1+\sqrt{3}\tan x\right)}}$.

Similarly, the last integral can be worked out to have $\dfrac{1}{3}\displaystyle\int\dfrac{dz}{\left(1-\sqrt{3}z\right)^{2}}=$

$\underline{\dfrac{\sqrt{3}}{9\left(1-\sqrt{3}\tan x\right)}}$. After collecting all underlined related answers

we get the solutions as $\dfrac{1}{6\sqrt{3}}\ln\left(\dfrac{1+\sqrt{3}\tan x}{1-\sqrt{3}\tan x}\right)-\dfrac{\sqrt{3}}{9\left(1+\sqrt{3}\tan x\right)}+$

$\dfrac{\sqrt{3}}{9\left(1-\sqrt{3}\tan x\right)}$. After some simplifications we get the answer as

$\dfrac{\sqrt{3}}{18}\left[\ln\left(\dfrac{1+\sqrt{3}\tan x}{1-\sqrt{3}\tan x}\right)+\dfrac{4}{1-3\tan^{2}x}\right].$

INTEGRAL 61

Problem:

$$\int \frac{\sin^3 x}{\sqrt{\cos x}}\,dx$$

Solution:

$$\frac{2\sqrt{\cos x}}{5}\left(\cos^2 x - 5\right) + constant$$

Techniques used:

Change of variables, Trigonometric identities

Step-by-step solution:

Having $\dfrac{d}{dx}\left(\sqrt{\cos x}\right) = \dfrac{-\sin x}{2\sqrt{\cos x}}$, rewrite the integral as

$\int \dfrac{\sin^3 x}{\sqrt{\cos x}}\,dx = -2\int\left(\dfrac{-\sin x}{2\sqrt{\cos x}}\,dx\right)\sin^2 x.$ Now, using the integra-

tion by parts technique, we get $-2\int\underbrace{\left(\dfrac{-\sin x}{2\sqrt{\cos x}}\,dx\right)}_{dg}\underbrace{\sin^2 x}_{f} =$

$-2\left(\sqrt{\cos x}\,\sin^2 x - 2\int\sqrt{\cos x}\,\sin x\cos x\,dx\right).$ But the latter integral

can be worked out as $-2\int\sqrt{\cos x}\,\sin x\cos x\,dx = -2\int\cos^{3/2} x\sin x\,dx =$

$\dfrac{4}{5}\cos^{5/2} x.$ After collecting all related terms, we get the answer as

$-2\left(\sqrt{\cos x}\,\sin^2 x + \dfrac{4}{5}\cos^{5/2} x\right).$ This expression can be simplified to

$-2\left(\sqrt{\cos x}\,\sin^2 x + \dfrac{4}{5}\cos^{\frac{5}{2}} x\right) = \dfrac{2\sqrt{\cos x}}{5}\left(-5\left(1 - \cos^2 x\right) - 4\cos^2 x\right) =$

$\dfrac{2\sqrt{\cos x}}{5}\left(\cos^2 x - 5\right).$

INTEGRAL 62

Problem:

$$\int \frac{5x+31}{3x^2-4x+11}\,dx$$

Solution:

$$\frac{5}{6}\ln\left(3x^2-4x+11\right)+\frac{103}{3\sqrt{29}}\tan^{-1}\left(\frac{3x-2}{\sqrt{29}}\right)+constant$$

Techniques used:

Change of variables, Integration by parts

Step-by-step solution:

Having $\dfrac{d}{dx}\left(3x^2-4x+11\right)=6x-4$, we can rewrite the numerator as $5x+31=\dfrac{5}{6}(6x-4)+\dfrac{103}{3}$. Therefore, the integral can be written as $\displaystyle\int\frac{5x+31}{3x^2-4x+11}\,dx=\frac{5}{6}\int\frac{6x-4}{3x^2-4x+11}\,dx+\frac{103}{3}\int\frac{1}{3x^2-4x+11}\,dx.$ But $\dfrac{5}{6}\displaystyle\int\frac{6x-4}{3x^2-4x+11}\,dx=\frac{5}{6}\ln\left(3x^2-4x+11\right).$ For calculating the latter integral, rewrite the denominator as $3x^2-4x+11=\left(\sqrt{3}x-\dfrac{2}{\sqrt{3}}\right)^2-\dfrac{4}{3}+11=\left(\sqrt{3}x-\dfrac{2}{\sqrt{3}}\right)^2+\dfrac{29}{3}.$ Now let $\sqrt{3}x-\dfrac{2}{\sqrt{3}}=z\Rightarrow dx=\dfrac{\sqrt{3}}{3}dz$ and the integral in terms of variable z reads $\dfrac{103}{3}\displaystyle\int\frac{1}{3x^2-4x+11}\,dx=\frac{103\sqrt{3}}{9}\int\frac{1}{\dfrac{29}{3}+z^2}\,dz=\frac{103}{29\sqrt{3}}$ $\displaystyle\int\frac{1}{1+\left(\sqrt{3/29}z\right)^2}\,dz.$ Now let $\sqrt{3/29}z=u\Rightarrow dz=\dfrac{\sqrt{29}}{\sqrt{3}}du,$ thus the integral in terms of variable u reads $\dfrac{103}{29\sqrt{3}}\displaystyle\int \dfrac{1}{1+\left(\sqrt{3/29}z\right)^2}\,dz =$

$\dfrac{103}{29\sqrt{3}}\dfrac{\sqrt{29}}{\sqrt{3}}\displaystyle\int \dfrac{1}{1+u^2}\,du = \dfrac{103}{3\sqrt{29}}\displaystyle\int \dfrac{1}{1+u^2}\,du.$ But $\displaystyle\int \dfrac{1}{1+u^2}\,du = \tan^{-1}u.$

Therefore, $\dfrac{103}{3\sqrt{29}}\displaystyle\int \dfrac{1}{1+u^2}\,du = \dfrac{103}{3\sqrt{29}}\tan^{-1}u = \dfrac{103}{3\sqrt{29}}\tan^{-1}\left(\sqrt{3/29}z\right)$

$= \dfrac{103}{3\sqrt{29}}\tan^{-1}\left(\sqrt{3/29}\left(\sqrt{3}x - \dfrac{2}{\sqrt{3}}\right)\right) = \dfrac{103}{3\sqrt{29}}\tan^{-1}\left(\dfrac{3x-2}{\sqrt{29}}\right).$

Collecting all related terms, we get the answer as $\dfrac{5}{6}\ln\left(3x^2 - 4x + 11\right) +$

$\dfrac{103}{3\sqrt{29}}\tan^{-1}\left(\dfrac{3x-2}{\sqrt{29}}\right).$

INTEGRAL 63

Problem:

$$\int \frac{3x^5 - x^4 + 2x^3 - 12x^2 - 2x + 1}{\left(x^3 - 1\right)^2} dx$$

Solution:

$$\ln\left(x^3 - 1\right) + \frac{x^2 - x + 3}{x^3 - 1} + constant$$

Techniques used:

Partial fractions, Change of variables, Integration by parts

Step-by-step solution:

Having the algebraic identity $x^3 - 1 = (x-1)(x^2 + x + 1)$, the denominator polynomial common factors are $(x-1)$, $(x-1)^2$, $(x^2 + x + 1)$, $(x^2 + x + 1)^2$. Using the partial fractions technique we form the following equation, which should be satisfied, for specific values of unknown polynomials $P(x) \equiv P, Q(x) \equiv Q, R(x) \equiv R, S(x) \equiv S$:

$$\frac{3x^5 - x^4 + 2x^3 - 12x^2 - 2x + 1}{\left(x^3 - 1\right)^2} = \frac{P}{x-1} + \frac{Q}{(x-1)^2} + \frac{R}{(x^2 + x + 1)} +$$

$$\frac{S}{\left(x^2 + x + 1\right)^2}.$$ After some manipulations (i.e., equating equal

power of x coefficients) the expressions for the unknown polynomials can be obtained as $P = -Q = 1, R = 2x + 1$, and $S = 4x + 2$. Therefore, the integral can be written as

$$\int \frac{3x^5 - x^4 + 2x^3 - 12x^2 - 2x + 1}{\left(x^3 - 1\right)^2} dx = \int \frac{1}{x-1} dx - \int \frac{1}{(x-1)^2} dx +$$

$$\int \frac{2x+1}{x^2+x+1}dx + 2\int \frac{2x+1}{\left(x^2+x+1\right)^2}dx. \quad \text{But} \quad \int \frac{1}{x-1}dx = \ln(x-1). \quad \text{For}$$

$$-\int \frac{1}{\left(x-1\right)^2}dx, \text{ let } x-1=z \Rightarrow dx = dz \text{ and rewriting the integral}$$

in terms of variable z gives $-\int \frac{1}{\left(x-1\right)^2}dx = -\int \frac{1}{z^2}dz = \frac{1}{z} = \frac{1}{x-1}.$

For $\int \frac{2x+1}{x^2+x+1}dx$, having $\frac{d}{dx}\left(x^2+x+1\right) = 2x+1$, we get

$$\int \frac{2x+1}{x^2+x+1}dx = \ln\left(x^2+x+1\right). \quad \text{For} \quad 2\int \frac{2x+1}{\left(x^2+x+1\right)^2}dx \quad \text{let}$$

$$x^2+x+1 = u \Rightarrow dx = \frac{du}{2x+1} \quad \text{and} \quad \text{rewrite} \quad \text{the} \quad \text{integral} \quad \text{as}$$

$$2\int \frac{2x+1}{\left(x^2+x+1\right)^2}dx = 2\int \frac{2x+1}{u^2}\frac{du}{2x+1} = 2\int \frac{1}{u^2}du = -\frac{2}{u} = -\frac{2}{x^2+x+1}.$$

After collecting all related terms, we get the answer as

$\ln(x-1) + \frac{1}{x-1} + \ln\left(x^2+x+1\right) - \frac{2}{x^2+x+1}.$ This expression can be

simplified to $\ln\left(x^3-1\right) + \frac{x^2-x+3}{x^3-1}.$

INTEGRAL 64

Problem:

$$\int \frac{4x^3 - x + 1}{x^3 + 1} dx$$

Solution:

$$4x + \frac{1}{3}\ln\left[\frac{x^2 - x + 1}{(x+1)^2}\right] - \frac{4}{\sqrt{3}}\tan^{-1}\left(\frac{2x - 1}{\sqrt{3}}\right) + constant$$

Techniques used:

Partial fractions, Change of variables, Integration by parts

Step-by-step solution:

Having the order of polynomials for both numerator and denominator of the integrand being equal, we apply division operation to get $\frac{4x^3 - x + 1}{x^3 + 1} = 4 - \frac{x + 3}{x^3 + 1}$. Therefore, the integral can be written as $\int \frac{4x^3 - x + 1}{x^3 + 1} dx = 4\int dx - \int \frac{x + 3}{x^3 + 1} dx$. But $4\int dx = 4x$ and for calculating the remaining integral, we use partial fractions method; $\frac{x + 3}{x^3 + 1} = \frac{x + 3}{(x+1)(x^2 - x + 1)} = \frac{P}{x + 1} + \frac{Q}{x^2 - x + 1}$. Therefore, to maintain the equality we must have $x + 3 \equiv P(x^2 - x + 1) + Q(x + 1)$ which requires to have $Q = Ax + B$. Therefore, after equating the coefficients of equal power of x on both sides of the equality, we have $\begin{cases} A + P = 0 \\ A + B - P = 1 \\ B + P = 3 \end{cases}$. The solution for this system of equations is $P = -A = 2/3$ and $B = 7/3$. Using these values, we   can   write $-\int \dfrac{x+3}{x^3+1}dx = -\dfrac{2}{3}\int \dfrac{1}{x+1}dx + \int \dfrac{2x-7}{3(x^2-x+1)}dx.$

But $= -\dfrac{2}{3}\int \dfrac{1}{x+1}dx = -\dfrac{2}{3}\ln(x+1).$ For the remaining integral,

rewrite it as $\int \dfrac{2x-7}{3(x^2-x+1)}dx = \dfrac{1}{3}\int \dfrac{2x-7}{x^2-x+1}dx = \dfrac{1}{3}\int \dfrac{2x-1-6}{x^2-x+1}dx =$

$\dfrac{1}{3}\int \dfrac{2x-1}{x^2-x+1}dx - 2\int \dfrac{1}{x^2-x+1}dx.$ But $\dfrac{1}{3}\int \dfrac{2x-1}{x^2-x+1}dx = \dfrac{1}{3}\ln$

$(x^2-x+1).$ Note that $\dfrac{d}{dx}(x^2-x+1) = 2x-1.$ Now for per-

forming integration for the remaining integral, rewrite it as

$-2\int \dfrac{1}{x^2-x+1}dx = -2\int \dfrac{1}{\left(x-\dfrac{1}{2}\right)^2 - \dfrac{1}{4}+1}dx = -2\int \dfrac{1}{\left(x-\dfrac{1}{2}\right)^2 + \dfrac{3}{4}}dx =$

$-\dfrac{8}{3}\int \dfrac{1}{\left(\dfrac{2x-1}{\sqrt{3}}\right)^2 + 1}dx.$ Now, let $\dfrac{2x-1}{\sqrt{3}} = z \Rightarrow dx = \dfrac{\sqrt{3}}{2}dz$ and we can

write the last integral in terms of the variable z as

$-\dfrac{8}{3}\int \dfrac{1}{\left(\dfrac{2x-1}{\sqrt{3}}\right)^2 + 1}dx = -\dfrac{4\sqrt{3}}{3}\int \dfrac{1}{z^2+1}dz.$ But $\int \dfrac{1}{z^2+1}dz = \tan^{-1}z.$

Therefore, $-\dfrac{4\sqrt{3}}{3}\int \dfrac{1}{z^2+1}dz = -\dfrac{4\sqrt{3}}{3}\tan^{-1}z = -\dfrac{4\sqrt{3}}{3}\tan^{-1}\left(\dfrac{2x-1}{\sqrt{3}}\right).$

Collecting all related terms, we arrive at the solution of the original

integral as $4x - \dfrac{2}{3}\ln(x+1) + \dfrac{1}{3}\ln(x^2-x+1) - \dfrac{4\sqrt{3}}{3}\tan^{-1}\left(\dfrac{2x-1}{\sqrt{3}}\right).$

INTEGRAL 65

Problem:

$$\int \sin^{-1} x \ln x \, dx$$

Solution:

$$\sqrt{1-x^2}\left(\ln x - 2\right) + x\sin^{-1} x\left(\ln x - 1\right) + \tanh^{-1}\left(\sqrt{1-x^2}\right) + constant$$

Techniques used:

Change of variables, Integration by parts, Trigonometric identities

Step-by-step solution:

Using integration by parts, we can write the integral as $\int \underset{dg}{\underline{\sin^{-1}}} x\underset{f}{\underline{\ln x}}\,dx = \ln x\int \sin^{-1} x\,dx - \int\left[\frac{dx}{x}\left(\int \sin^{-1} x\,dx\right)\right]$. Therefore, we require to have $\int \sin^{-1} x\,dx$. Let $\sin^{-1} x = \alpha \Rightarrow \sin\alpha = x$ and $\cos\alpha\,d\alpha = dx$. Therefore, we can write $\int \sin^{-1} x\,dx = \int \alpha\cos\alpha\,d\alpha = \alpha\sin\alpha - \underbrace{\int \sin\alpha\,d\alpha}_{-\cos\alpha}$. In terms of the original variable x, we get $\int \sin^{-1} x\,dx = x\sin^{-1} x + \sqrt{1-x^2}$. Now, back to the original problem, we can write $\int \sin^{-1} x\ln x\,dx = \ln x\int \sin^{-1} xdx - \int\left[\frac{dx}{x}\right.$ $\left.\left(\int \sin^{-1} x\,dx\right)\right] = \ln x\left(x\sin^{-1} x + \sqrt{1-x^2}\right) - \int \frac{1}{x}\left(x\sin^{-1} x + \sqrt{1-x^2}\right)dx$.

Expanding the remaining integral, we get $\int \frac{1}{x}\left(x\sin^{-1} x + \sqrt{1-x^2}\right)dx$

$= \int \sin^{-1} x\,dx + \int \frac{\sqrt{1-x^2}}{x}\,dx = x\sin^{-1} x + \sqrt{1-x^2} + \int \frac{\sqrt{1-x^2}}{x}\,dx.$

For the latter integral, let $x = \sin z \Rightarrow dx = \cos z\,dz$. Rewriting the integral in terms of variable z, gives $\int \dfrac{\sqrt{1-x^2}}{x} dx =$

$\int \dfrac{\sqrt{1-\sin^2 z}}{\sin z} \cos z\, dz = \int \dfrac{\cos^2 z}{\sin z} dz$. But $\int \dfrac{\cos^2 z}{\sin z} dz = \int \dfrac{1-\sin^2 z}{\sin z} dz =$

$\int \dfrac{1}{\sin z} dz - \int \sin z\, dz$. For the former integral, let $\cos z = u \Rightarrow -$

$\underset{-\cos z}{\underbrace{}}$

$\sin z\, dz = du$ and rewrite the integral in terms of varia-

ble u as $\int \dfrac{1}{\sin z} dz = \int \dfrac{1}{\sin^2 z} du - -\int \dfrac{1}{1-u^2} du$. But we have

$\int \dfrac{1}{1-u^2} du = \tanh^{-1} u$ and rewriting in terms of the original

variable x, we get $\tanh^{-1} u = \tanh^{-1}(\cos z) = \tanh^{-1}\left(\sqrt{1-x^2}\right)$.

After collecting all related terms, we get the answer

as $\ln x\left(x\sin^{-1} x + \sqrt{1-x^2}\right) - \left(x\sin^{-1} x + \sqrt{1-x^2}\right) - \sqrt{1-x^2} + \tanh^{-1}$

$\left(\sqrt{1-x^2}\right) + \sqrt{1-x^2}$. This expression simplifies to $\sqrt{1-x^2}\left(\ln x - 2\right) +$

$x\sin^{-1} x\left(\ln x - 1\right) + \tanh^{-1}\left(\sqrt{1-x^2}\right) + \sqrt{1-x^2}$.

INTEGRAL 66

Problem:

$$\int \ln(\sin x)\sqrt{1+\sin x}\,dx$$

Solution:

$$\left(\sin\frac{x}{2}-\cos\frac{x}{2}\right)\ln 4 + 2\left\{\left(\sin\frac{x}{2}-\cos\frac{x}{2}\right)\left[\ln\left(\sin\frac{x}{2}\right)+\ln\left(\cos\frac{x}{2}\right)-2\right]\right\}$$

$$+2\left[\tanh^{-1}\left(\sin\frac{x}{2}\right)-\tanh^{-1}\left(\cos\frac{x}{2}\right)\right]+constant$$

Techniques used:

Change of variables, Integration by parts, Trigonometric identities

Step-by-step solution:

After writing the trigonometric functions in half-angle form, we get

$$\int \ln(\sin x)\sqrt{1+\sin x}\,dx = \int \ln\left(2\sin\frac{x}{2}\cos\frac{x}{2}\right)\sqrt{\sin^2\frac{x}{2}+\cos^2\frac{x}{2}+2\sin\frac{x}{2}}$$

$$\cos\frac{x}{2}\,dx = \int \ln\left(2\sin\frac{x}{2}\cos\frac{x}{2}\right)\sqrt{\left(\sin\frac{x}{2}\cos\frac{x}{2}\right)^2}\,dx = \int \ln\left(2\sin\frac{x}{2}\cos\frac{x}{2}\right)$$

$$\left(\sin\frac{x}{2}\cos\frac{x}{2}\right)dx.$$ Expanding the logarithm expression, we get

$$\ln\left(2\sin\frac{x}{2}\cos\frac{x}{2}\right)=\left(\ln 2 + \ln\sin\frac{x}{2}+\ln\cos\frac{x}{2}\right).$$ Therefore, we can

write the integral in terms of summation of its parts, as

$$\ln 2\left(\int\sin\frac{x}{2}\,dx+\int\cos\frac{x}{2}\,dx\right),\quad \int\sin\frac{x}{2}\ln\sin\frac{x}{2}\,dx,\quad \int\cos\frac{x}{2}\ln\sin\frac{x}{2}\,dx,$$

$$\int\sin\frac{x}{2}\ln\cos\frac{x}{2}\,dx,\int\cos\frac{x}{2}\ln\cos\frac{x}{2}\,dx.$$ Now we perform the integration

operation for new integrals. Therefore, $\int \sin\dfrac{x}{2}dx = -2\cos\dfrac{x}{2}$,

$\int \cos\dfrac{x}{2}dx = 2\sin\dfrac{x}{2}$. Using the integration by parts technique,

we have $\int \sin\dfrac{x}{2}\ln\sin\dfrac{x}{2}dx = -2\cos\dfrac{x}{2}\ln\sin\dfrac{x}{2} + \int\dfrac{\cos^2\dfrac{x}{2}}{\sin\dfrac{x}{2}}dx$. But

$\int\dfrac{\cos^2\dfrac{x}{2}}{\sin\dfrac{x}{2}}dx = \int\dfrac{1-\sin^2\dfrac{x}{2}}{\sin\dfrac{x}{2}}dx = \int\dfrac{1}{\sin\dfrac{x}{2}}dx - \underbrace{\int\sin\dfrac{x}{2}dx}_{=2\cos\frac{x}{2}}$. But for $\int\dfrac{1}{\sin\dfrac{x}{2}}dx$,

let $\cos\dfrac{x}{2} = z \Rightarrow -\dfrac{1}{2}\sin\dfrac{x}{2}dx = dz$ and rewrite the integral in terms

of variable z, we get $\int\dfrac{1}{\sin\dfrac{x}{2}}dx = -2\int\dfrac{1}{\sin^2\dfrac{x}{2}}dz = -2\int\dfrac{1}{1-z^2}dz =$

$-2\tanh^{-1}z = -2\tanh^{-1}\left(\cos\dfrac{x}{2}\right)$. Similarly, we can perform the integration process for the remaining integrals. These results are tabulated as shown below:

Integral	Answer
$\int \sin\dfrac{x}{2}\ln\sin\dfrac{x}{2}dx$	$2\left[-\cos\dfrac{x}{2}\ln\sin\dfrac{x}{2} + \cos\dfrac{x}{2} - \tanh^{-1}\left(\cos\dfrac{x}{2}\right)\right]$
$\int \cos\dfrac{x}{2}\ln\sin\dfrac{x}{2}dx$	$2\left(\sin\dfrac{x}{2}\ln\sin\dfrac{x}{2} - \sin\dfrac{x}{2}\right)$
$\int \sin\dfrac{x}{2}\ln\cos\dfrac{x}{2}dx$	$2\left(-\cos\dfrac{x}{2}\ln\cos\dfrac{x}{2} + \cos\dfrac{x}{2}\right)$
$\int \cos\dfrac{x}{2}\ln\cos\dfrac{x}{2}dx$	$2\left[\sin\dfrac{x}{2}\ln\cos\dfrac{x}{2} - \sin\dfrac{x}{2} + \tanh^{-1}\left(\sin\dfrac{x}{2}\right)\right]$

After collecting all related terms and simplifying, we receive the answer as shown above, as solution.

INTEGRAL 67

Problem:

$$\int \frac{x^3 e^{\sin^{-1} x}}{\sqrt{1-x^2}}\, dx$$

Solution:

$$\frac{3e^{\sin^{-1} x}}{10}\left[x + \frac{x^3}{3} - \left(1+x^2\right)\sqrt{1-x^2} \right] + constant$$

Techniques used:

Change of variables, Integration by parts, Trigonometric identities

Step-by-step solution:

Having $\dfrac{d}{dx}\left(\sin^{-1} x\right) = \dfrac{1}{\sqrt{1-x^2}}$, use the integration by parts technique

to get $\displaystyle\int \frac{x^3 e^{\sin^{-1} x}}{\sqrt{1-x^2}}\, dx = \int \underbrace{x^3}_{f}\, e^{\sin^{-1} x}\, \underbrace{\frac{dx}{\sqrt{1-x^2}}}_{dg} = x^3 e^{\sin^{-1} x} - 3\int x^2 e^{\sin^{-1} x}\, dx.$ For

the new integral, let $\sin^{-1} x = z \Rightarrow \sin z = x, dx = \cos z\, dz,$ and write the

integral in terms of variable z as $-3\int x^2 e^{\sin^{-1} x}\, dx = -3\int \sin^2 z \cos z\, e^z\, dz.$

After applying the integration by parts technique, we get

$-3\int \sin^2 z \cos z\, e^z\, dz = -3\left(\dfrac{\sin^3 z}{3} \right) e^z + \int \sin^3 z\, e^z\, dz.$ For the new

integral, after substituting $\sin^3 z = \dfrac{3}{4}\sin z - \dfrac{1}{4}\sin(3z),$ we have

$\displaystyle\int \sin^3 z\, e^z\, dz = \frac{3}{4}\int e^z \sin z\, dz - \frac{1}{4}\int e^z \sin(3z)\, dz.$ Now we have two

integrals to calculate. For $\int e^z \sin z\, dz,$ apply the integration by

parts technique twice to get $\displaystyle\int e^z \sin z\, dz = -e^z \cos z + \int e^z \cos z\, dz = -$

$e^z \cos z + e^z \sin z - \int e^z \sin z$. Now, after rearranging the terms and rewriting this expression as $2\int e^z \sin z \, dz = -e^z \cos z + e^z \sin z$, we get

$\int e^z \sin z \, dz = \dfrac{e^z}{2}(\sin z - \cos z)$. Or in terms of the original variable x we

have $\dfrac{e^z}{2}(\sin z - \cos z) = \dfrac{e^{\sin^{-1} x}}{2}\left(x - \sqrt{1 - x^2}\right)$. Similarly, apply the same

integration process to the remaining integral, $\int e^z \sin(3z) \, dz$. Therefore,

we get $\int e^z \sin(3z) \, dz = e^z\left(-\dfrac{1}{3}\cos 3z\right) + \dfrac{1}{3}\int e^z \cos 3z \, dz = \dfrac{-e^z}{3}(\cos 3z)$

$+ \dfrac{1}{3}\left(\dfrac{e^z}{3}\sin 3z - \dfrac{1}{3}\int e^z \sin 3z \, dz\right)$. Now, after rearranging the terms and

writing this expression, we get $\int e^z \sin(3z) \, dz = \dfrac{e^z}{10}(\sin 3z - 3\cos 3z)$.

Or in terms of the original variable x we have $\dfrac{e^z}{10}(\sin 3z - 3\cos 3z) =$

$\dfrac{e^{\sin^{-1} x}}{10}\left(\sin\left(3\sin^{-1} x\right) - 3\cos\left(3\sin^{-1} x\right)\right)$. Collecting all related terms

gives the answer as $\dfrac{3}{8}e^{\sin^{-1} x}\left(x - \sqrt{1 - x^2}\right) - \dfrac{1}{40}e^{\sin^{-1} x}\left[\sin\left(3\sin^{-1} x\right) - \right.$

$\left. 3\cos\left(3\sin^{-1} x\right)\right]$. Using the expansion of the trigonometric functions,

we can simplify the answer as $\dfrac{3}{8}e^{\sin^{-1} x}\left(x - \sqrt{1 - x^2}\right) - \dfrac{1}{40}e^{\sin^{-1} x}$

$\left[-x^3 - 3\left(1 - x^2\right)\sqrt{1 - x^2} + 3x\left(1 - x^2\right) + 9x^2\sqrt{1 - x^2}\right] = \dfrac{3e^{\sin^{-1} x}}{10}$

$\left[x + \dfrac{1}{3}x^3 - \sqrt{1 - x^2}\left(1 + x^2\right)\right]$.

INTEGRAL 68

Problem:

$$\int \frac{\tan x}{\sqrt{1+\sec^3 x}}\,dx$$

Solution:

$$-\frac{2}{3}\tanh^{-1}\sqrt{1+\sec^3 x}+constant$$

Techniques used:

Change of variables, Integration by parts, Trigonometric identities

Step-by-step solution:

Rewrite the integrand as $\dfrac{\tan x}{\sqrt{1+\sec^3 x}}=\dfrac{\sin x/\cos x}{\sqrt{1+1/\cos^3 x}}=\dfrac{\sin x}{\sqrt{\dfrac{1+\cos^3 x}{\cos x}}}.$

Let $\dfrac{1}{\cos x}=z\Rightarrow dx=\dfrac{dz}{z^2\sin x}$ and write the integral in terms of

the variable z to get $\displaystyle\int\frac{\sin x}{\sqrt{\dfrac{1+\cos^3 x}{\cos x}}}\,dx=\int\frac{\sin x}{\sqrt{z+1/z^2}}\frac{dz}{z^2\sin x}=$

$\displaystyle\int\frac{1}{z\sqrt{1+z^3}}\,dz.$ Now let $1+z^3=u\Rightarrow dz=\dfrac{du}{3z^2}.$ Therefore, we

have $\displaystyle\int\frac{1}{z\sqrt{1+z^3}}\,dz=\frac{1}{3}\int\frac{1}{(u-1)\sqrt{u}}\,du.$ Let $\sqrt{u}=y\Rightarrow du=2ydy$

and rewrite the latter integral in terms of the variable y as,

$\dfrac{1}{3}\displaystyle\int\frac{1}{(u-1)\sqrt{u}}\,du=\frac{2}{3}\int\frac{y}{(y^2-1)y}\,dy=-\frac{2}{3}\int\frac{1}{1-y^2}\,dy=-\frac{2}{3}\tanh^{-1}y.$

After several substitutions, we can write the answer in terms of

the original variable x as $-\dfrac{2}{3}\tanh^{-1}y=-\dfrac{2}{3}\tanh^{-1}\sqrt{u}=-\dfrac{2}{3}\tanh^{-1}$

$\sqrt{1+z^3}=-\dfrac{2}{3}\tanh^{-1}\sqrt{1+\sec^3 x}.$

INTEGRAL 69

Problem:

$$\int \frac{x\ln\left(x+\sqrt{x^2-1}\right)}{\sqrt{x^2-1}}dx$$

Solution:

$$\sqrt{x^2-1}\ln\left(x+\sqrt{x^2-1}\right)-x+constant$$

Techniques used:

Change of variables, Integration by parts

Step-by-step solution:

Having $\dfrac{d}{dx}\left(x+\sqrt{x^2-1}\right)=1+\dfrac{x}{\sqrt{x^2-1}}=\dfrac{x+\sqrt{x^2-1}}{\sqrt{x^2-1}}$ and rewriting

the $\dfrac{x}{\sqrt{x^2-1}}=\underset{0}{\dfrac{x}{\sqrt{x^2-1}}+1-1}$, we have the integral as follows:

$$\int \frac{x\ln\left(x+\sqrt{x^2-1}\right)}{\sqrt{x^2-1}}dx=\int\left(\frac{x}{\sqrt{x^2-1}}+1-1\right)\ln\left(x+\sqrt{x^2-1}\right)dx=$$

$$\int\left(1+\frac{x}{\sqrt{x^2-1}}\right)\ln\left(x+\sqrt{x^2-1}\right)dx-\int\ln\left(x+\sqrt{x^2-1}\right)dx.$$

Now we have two integrals to calculate. For the latter integral, use the integration by parts technique to get

$$-\int\ln\left(x+\sqrt{x^2-1}\right)dx=-x\ln\left(x+\sqrt{x^2-1}\right)+\int x\frac{\left(1+\dfrac{x}{\sqrt{x^2-1}}\right)}{x+\sqrt{x^2-1}}dx=$$

$$-x\ln\left(x+\sqrt{x^2-1}\right)+\int\frac{x}{\sqrt{x^2-1}}dx.\ \text{Now let }x^2-1=z\Rightarrow xdx=\frac{dz}{2}$$

and write the remaining integral in terms of variable z

as $\int\dfrac{x}{\sqrt{x^2-1}}dx=\dfrac{1}{2}\int\dfrac{1}{\sqrt{z}}dz=\sqrt{z}$, or $\sqrt{x^2-1}$. For the former integral, let $x + \sqrt{x^2 - 1} = u \Rightarrow \left(1 + \dfrac{x}{\sqrt{x^2 - 1}}\right) dx = du.$ Hence

$\int \left(1 + \dfrac{x}{\sqrt{x^2 - 1}}\right) \ln\left(x + \sqrt{x^2 - 1}\right) dx = \int \ln u \, du = u \ln u - u.$

Therefore, in terms of the variable x, we get $u \ln u - u =$
$\left(x + \sqrt{x^2 - 1}\right)\left(\ln\left(x + \sqrt{x^2 - 1}\right) - 1\right).$ Collecting all related terms gives

$-x \ln\left(x + \sqrt{x^2 - 1}\right) + \sqrt{x^2 - 1} + \left(x + \sqrt{x^2 - 1}\right)\left(\ln\left(x + \sqrt{x^2 - 1}\right) - 1\right) =$

$\sqrt{x^2 - 1} \ln\left(x + \sqrt{x^2 - 1}\right) - x.$

INTEGRAL 70

Problem:

$$\int \sin^{-1}\left(\sqrt{1+x}\right)dx$$

Solution:

$$\frac{1}{2}\left[(1+2x)\sin^{-1}\left(\sqrt{1+x}\right)+\sqrt{-x(1+x)}\right]+constant$$

Techniques used:

Change of variables, Integration by parts, Trigonometric identities

Step-by-step solution:

Let $\sqrt{1+x}=z\Rightarrow dx=2zdz$ and write the integral in terms of the variable z to get $\int \sin^{-1}\left(\sqrt{1+x}\right)dx=2\int z\sin^{-1}zdz$. Using

the integration by parts technique, we have $2\int \underbrace{z}_{dg}\underbrace{\sin^{-1}z}_{f}dz=z^2$

$\sin^{-1}z-\int \dfrac{z^2}{\sqrt{1-z^2}}dz$. Note that $\dfrac{d}{dz}\left(\sin^{-1}z\right)=\dfrac{1}{\sqrt{1-z^2}}$. Therefore,

$z^2\sin^{-1}z=\underline{(1+x)\sin^{-1}\sqrt{1+x}}$. To calculate the remaining integral,

rewrite it as $-\int \dfrac{z^2}{\sqrt{1-z^2}}dz=\int z\dfrac{-z}{\sqrt{1-z^2}}dz$ and apply the integration

by parts technique to get $\int \underbrace{z}_{f}\underbrace{\dfrac{(-z)}{\sqrt{1-z^2}}}_{dg}dz=z\sqrt{1-z^2}-\int \sqrt{1-z^2}dz$.

Therefore, $z\sqrt{1-z^2}=\sqrt{-x(1+x)}$. To calculate the latter integral,

let $z=\sin u\Rightarrow dz=\cos u\,du$ and write the integral in terms of the

variable u to get $-\int \sqrt{1-z^2}dz=-\int \cos^2 u\,du$. But $\cos^2 u=\dfrac{1+\cos 2u}{2}$.

Hence, $-\int \cos^2 u \, du = -\frac{1}{2}\int(1+\cos 2u)\,du = -\frac{u}{2} - \frac{\sin 2u}{4}$. Therefore,

$-\frac{u}{2} - \frac{\sin 2u}{4} = -\frac{1}{2}(u + \sin u \cos u) = -\frac{1}{2}\left(\sin^{-1} z + z\sqrt{1-z^2}\right) =$

$-\frac{1}{2}\left(\sin^{-1}\sqrt{1+x} + \sqrt{-x(1+x)}\right)$. After collecting all obtained results,

as underlined, and simplifying we get the solution as $\frac{1}{2}\Big[(1+2x)\sin^{-1}$

$\sqrt{1+x} + \sqrt{-x(1+x)}\Big]$.

INTEGRAL 71

Problem:

$$\int \frac{1}{2 + 2\sin x + \cos x} dx$$

Solution:

$$\ln\left[\frac{\tan(x/2)+1}{\tan(x/2)+3}\right] + constant$$

Techniques used:

Change of variables, Trigonometric identities, Partial fractions

Step-by-step solution:

The integral can be written in terms of half-angle using the trigo-nometric identities $\sin x = \dfrac{2\tan x/2}{1+\tan^2 x/2}$ and $\cos x = \dfrac{1-\tan^2 x/2}{1+\tan^2 x/2}$ as

$$\int \frac{1}{2+2\sin x + \cos x} dx = \int \frac{1}{2 + \dfrac{4\tan x/2}{1+\tan^2 x/2} + \dfrac{1-\tan^2 x/2}{1+\tan^2 x/2}} dx =$$

$$\int \frac{1}{\tan^2 x/2 + 4\tan x/2 + 3} \frac{dx}{\cos^2 x/2}. \quad \text{Now let } \tan\frac{x}{2} = z \Rightarrow dx =$$

$2\cos^2\dfrac{x}{2}dz$ and write the integral in terms of the variable z

as $2\int \dfrac{1}{z^2+4z+3}dz$. Using the partial fraction technique, we

get $2\int \dfrac{1}{z^2+4z+3}dz = 2\int \dfrac{1}{(z+3)(z+1)}dz = \int \dfrac{1}{z+1}dz - \int \dfrac{1}{z+3}dz$.

Performing the integration for the new integrals gives

$\int \dfrac{1}{z+1}dz = \ln(z+1)$ and $\int \dfrac{1}{z+3}dz = \ln(z+3)$. Collecting results

gives the answer as $\ln\dfrac{z+1}{z+3}$. Rewriting the answer in terms of the

original variable x gives $\ln\dfrac{z+1}{z+3} = \ln\left(\dfrac{\tan x/2+1}{\tan x/2+3}\right)$.

INTEGRAL 72

Problem:

$$\int \frac{\sec^2 x}{\tan^2 x + 2\tan x + 2}\,dx$$

Solution:

$$\tan^{-1}(1 + \tan x) + constant$$

Techniques used:

Change of variables, Integration by parts

Step-by-step solution:

Rewrite the denominator as $\tan^2 x + 2\tan x + 2 = 1 + (1 + \tan x)^2$.
Now let $(1 + \tan x) = z \Rightarrow dx \sec^2 x = dz$ and write the integral in
terms of the variable z to get $\int \dfrac{\sec^2 x}{\tan^2 x + 2\tan x + 2}\,dx = \int \dfrac{1}{1 + z^2}\,dz$.
But $\int \dfrac{1}{1+z^2}\,dz = \tan^{-1} z$, and in terms of the original variable x the
answer reads $\tan^{-1} z = \tan^{-1}(1 + \tan x)$.

INTEGRAL 73

Problem:

$$\int \sqrt{x^2 + x + 1}\, dx$$

Solution:

$$\frac{2x+1}{4}\sqrt{x^2 + x + 1} + \frac{3}{8}\tanh^{-1}\left(\frac{2x+1}{2\sqrt{x^2 + x + 1}}\right) + constant$$

Techniques used:

Change of variables, Integration by parts, Trigonometric identities

Step-by-step solution:

The expression $x^2 + x + 1$ can be written as $x^2 + x + 1 = \frac{3}{4} + \left(x + \frac{1}{2}\right)^2 = \frac{3}{4}\left[1 + \left(\frac{2x+1}{\sqrt{3}}\right)^2\right]$. Now, let $\frac{2x+1}{\sqrt{3}} = z \Rightarrow dx = \frac{\sqrt{3}}{2}dz$ and rewrite the integral in terms of variable z, as $\int \sqrt{x^2 + x + 1}\, dx =$

$\int\sqrt{\frac{3}{4}\left[1 + \left(\frac{2x+1}{\sqrt{3}}\right)^2\right]}dx = \frac{3}{4}\int\sqrt{1+z^2}\, dz$. For calculating the new integral, let $z = \tan u \Rightarrow dz = \frac{du}{\cos^2 u}$ and rewrite the integral in terms of variable u, as $\frac{3}{4}\int\sqrt{1+z^2}\, dz = \frac{3}{4}\int\sqrt{1+\tan^2 u}\,\frac{du}{\cos^2 u} = \frac{3}{4}\int\frac{1}{\cos^3 u}du$.

Using the integration by parts technique, we have $\frac{3}{4}\int\frac{1}{\cos^3 u}du = \frac{3}{4}\int\underbrace{\frac{du}{\cos^2 u}}_{dg}\underbrace{\frac{1}{\cos u}}_{f} = \frac{3}{4}\left(\frac{\tan u}{\cos u} - \int\frac{\tan u\sin u}{\cos^2 u}du\right)$. But

$\frac{\tan u\sin u}{\cos^2 u} = \frac{\sin^2 u}{\cos^3 u} = \frac{1-\cos^2 u}{\cos^3 u}$. Therefore, $\frac{3}{4}\int\frac{1}{\cos^3 u}du = \frac{3}{4}\left(\frac{\tan u}{\cos u}\right)$

$\frac{3}{4}\int\frac{1}{\cos^3 u}\,du+\frac{3}{4}\int\frac{1}{\cos u}\,du.$ Rearranging terms involved in the

last equation; we get $\frac{3}{4}\int\frac{1}{\cos^3 u}\,du=\frac{3}{8}\left(\frac{\tan u}{\cos u}\right)+\frac{3}{8}\int\frac{1}{\cos u}\,du.$

The solution to the new integral was presented in the Integral 5

section. Or $\frac{3}{8}\int\frac{1}{\cos u}\,du=\frac{3}{8}\tan^{-1}(\sin u)$. After collecting all related

terms, we get the answer in terms of the original variable x as

$$\frac{3}{8}\left(\frac{\tan u}{\cos u}\right)+\frac{3}{8}\tan^{-1}(\sin u)=\frac{3}{8}z\left(\sqrt{1+z^2}\right)+\frac{3}{8}\tanh^{-1}\left(\frac{z}{\sqrt{1+z^2}}\right)=$$

$$\frac{3}{8}\left[\frac{2x+1}{\sqrt{3}}\left(\sqrt{1+\left(\frac{2x+1}{\sqrt{3}}\right)^2}\right)\right]+\frac{3}{8}\tanh^{-1}\left(\frac{\frac{2x+1}{\sqrt{3}}}{\sqrt{1+\left(\frac{2x+1}{\sqrt{3}}\right)^2}}\right).$$ This

expression simplifies to $\dfrac{2x+1}{4}\sqrt{x^2+x+1}+\dfrac{3}{8}\tanh^{-1}\left(\dfrac{2x+1}{2\sqrt{x^2+x+1}}\right).$

INTEGRAL 74

Problem:

$$\int \frac{x}{\left(x^2 + 2x + 2\right)^2} \, dx$$

Solution:

$$-\frac{1}{2}\left[\tan^{-1}(x+1) + \frac{x+2}{x^2 + 2x + 2} \right] + constant$$

Techniques used:

Change of variables, Integration by parts, Trigonometric identities

Step-by-step solution:

Let $x+1 = z \Rightarrow dx = dz$ and rewrite the integral in terms of the variable z, to get $\int \frac{x}{\left(x^2 + 2x + 2\right)^2} \, dx = \int \frac{z-1}{\left[(z-1)^2 + 2(z-1) + 2\right]^2} \, dz =$

$\int \frac{z-1}{\left(1 + z^2\right)^2} \, dz$. The new integral can be written as $\int \frac{z-1}{\left(1 + z^2\right)^2} \, dz =$

$\int \frac{z}{\left(1 + z^2\right)^2} \, dz - \int \frac{1}{\left(1 + z^2\right)^2} \, dz$. Now we have two integrals to

calculate. For the first integral we can write $\int \frac{z}{\left(1 + z^2\right)^2} \, dz =$

$\frac{1}{2} \int \frac{2z}{\left(1 + z^2\right)^2} \, dz = -\frac{1}{2\left(1 + z^2\right)}$. For the remaining integral, let

$z = \tan u \Rightarrow dz = \frac{du}{\cos^2 u}$. Therefore, we have $\int \frac{1}{\left(1 + z^2\right)^2} \, dz =$

$\int \frac{1}{\left(1 + \tan^2 u\right)^2} \frac{du}{\cos^2 u} = \int \cos^2 u \, du$. But $\cos^2 u = \frac{1 + \cos 2u}{2}$ and the new integral reads $\int \cos^2 u\, du = \frac{1}{2}\int(1+\cos 2u)\,du = \frac{u}{2} + \frac{\sin 2u}{4}$. After collecting all related answers, in terms of the original variable x,

we have $-\dfrac{1}{2(1+z^2)} - \dfrac{u}{2} - \dfrac{\sin 2u}{4} = -\dfrac{1}{2(x^2+2x+2)} - \dfrac{\tan^{-1}(x+1)}{2} -$

$\dfrac{\sin(2\tan^{-1}(x+1))}{4}$. The answer can be simplified to,

$$\int \frac{x}{\left(x^2+2x+2\right)^2}\,dx = -\frac{\tan^{-1}(x+1)}{2} - \frac{x+2}{2\left(x^2+2x+2\right)}.$$

INTEGRAL 75

Problem:

$$\int \frac{1}{x^3 \sqrt{x^2 - 9}} dx$$

Solution:

$$\frac{1}{54}\left[\cos^{-1}\left(\frac{3}{x}\right) + \frac{3}{x^2} \sqrt{x^2 - 9} \right] + constant$$

Techniques used:

Change of variables, Trigonometric identities

Step-by-step solution:

Let $x = \dfrac{3}{\cos z} \Rightarrow dx = 3\dfrac{\sin z}{\cos^2 z} dz$ and rewrite the integral in terms of the variable z to get, $\dfrac{1}{9}\int \dfrac{\cos^3 z}{\sqrt{9/\cos^2 z - 9}} \dfrac{\sin z}{\cos^2 z} dz =$

$\dfrac{1}{27}\int \dfrac{\sin z \cos^2 z}{\sqrt{1 - \cos^2 z}} dz = \dfrac{1}{27}\int \cos^2 z dz$. Using trigonometric identity,

$\cos^2 z = \dfrac{1 + \cos 2z}{2}$, we get $\dfrac{1}{27}\int \cos^2 z dz = \dfrac{1}{54}\int (1 + \cos 2z) dz =$

$\dfrac{1}{54}\left(z + \dfrac{1}{2}\sin 2z \right)$. Now in terms of the original variable x we have the

answer as $\dfrac{1}{54}\left(z + \dfrac{1}{2}\sin 2z \right) = \dfrac{1}{54}(z + \sin z \cos z) = \dfrac{1}{54}\left[\cos^{-1}\left(\dfrac{3}{x}\right) + \right.$

$\dfrac{3}{x}\sqrt{1 - \dfrac{9}{x^2}}\left. \right]$. This expression simplifies to $\dfrac{1}{54}\left[\cos^{-1}\left(\dfrac{3}{x}\right) + \dfrac{3}{x^2} \right.$

$\sqrt{x^2 - 9}\left. \right]$, assuming $x > 0$. This expression simplifies to

$\dfrac{1}{54}\left[\cos^{-1}\left(\dfrac{3}{x}\right) + \dfrac{3}{x^2}\sqrt{x^2 - 9} \right]$.

INTEGRAL 76

Problem:

$$\int \sin x \tan^{-1} \sqrt{\sec x - 1}\, dx$$

Solution:

$$\frac{1}{2}\left(\cos^{-1}\sqrt{\cos x} + \sqrt{\cos x(1 - \cos x)}\right) - \cos x \tan^{-1}\sqrt{\sec x - 1} + constant$$

Techniques used:

Change of variables, Integration by parts, Trigonometric identities

Step-by-step solution:

Using the integration by parts technique we can write $\int \underset{dg}{\underbrace{\sin x}} \underset{f}{\underbrace{\tan^{-1} \sqrt{\sec x - 1}}}\, dx = -\cos x \tan^{-1} \sqrt{\sec x - 1} + \int \cos x \frac{d}{dx}\Big(\tan^{-1}$

$\sqrt{\sec x - 1}\Big) dx.$ But letting $\tan^{-1}\sqrt{\sec x - 1} = \alpha$, or $\tan \alpha = \sqrt{\sec x - 1}$,

we can write, by differentiating both sides, $\frac{d\alpha}{\cos^2 \alpha} = \frac{\sin x}{2\cos^2 x \sqrt{\sec x - 1}}\, dx.$ Now using $\tan \alpha = \sqrt{\sec x - 1}$, we

have $\cos^2 \alpha = \cos x$ and $d\alpha = \left(\tan^{-1}\sqrt{\sec x - 1}\right)' = \frac{\sin x\, dx}{2\cos x \sqrt{\sec x - 1}}.$

Therefore, $\int \cos x \underset{d\alpha}{\underbrace{\frac{\sin x}{2\cos x \sqrt{\sec x - 1}}}}\, dx = \frac{1}{2}\int \frac{\sin x}{\sqrt{\sec x - 1}}\, dx.$ For cal-

culating the latter integral, let $\cos x = \cos^2 z \Rightarrow \sin x\, dx = 2 \sin z$

$\cos z\, dz$ and write the integral in terms of the variable z as

$\frac{1}{2}\int \frac{\sin x}{\sqrt{\sec x - 1}}\, dx = \int \frac{\sin z \cos z}{\sqrt{1/\cos^2 z - 1}}\, dz = \int \cos^2 z\, dz.$ Now using the trigonometric identity $\cos^2 z = \dfrac{1+\cos 2z}{2}$, we have $\int \cos^2 z \, dz =$

$\dfrac{1}{2}\int (1+\cos 2z)\,dz = \dfrac{z}{2}+\dfrac{\sin 2z}{4}$. Therefore, in terms of the orig-

inal variable x, we can write $\dfrac{z}{2}+\dfrac{\sin 2z}{4}=\dfrac{1}{2}(z+\sin z \cos z) =$

$\dfrac{1}{2}\left(\cos^{-1}\sqrt{\cos x}+\sqrt{\cos x (1-\cos x)}\right)$. Collecting all related terms,

we receive the answer as $\dfrac{1}{2}\left(\cos^{-1}\sqrt{\cos x}+\sqrt{\cos x (1-\cos x)}\right) -$

$\cos x \tan^{-1}\sqrt{\sec x - 1}$.

INTEGRAL 77

Problem:

$$\int \frac{1}{\sqrt{x\sqrt{x} - x^2}}\, dx$$

Solution:

$$2\sin^{-1}\left(2\sqrt{x} - 1\right) + constant$$

Techniques used:

Change of variables, Integration by parts

Step-by-step solution:

Let $\sqrt{x} = z \Rightarrow dx = 2z\, dz$, and write the integral in terms of the variable z as $\int \frac{1}{\sqrt{x\sqrt{x} - x^2}}\, dx = 2\int \frac{z}{\sqrt{z^3 - z^4}}\, dz = 2\int \frac{1}{\sqrt{z - z^2}}\, dz$. Write the denominator as $\sqrt{z - z^2} = \sqrt{1/4 - (z - 1/2)^2} = \frac{1}{2}\sqrt{1 - (2z - 1)^2}$.

Now let $(2z - 1) = u \Rightarrow 2dz = du$ and write the new integral in terms of the variable u as $2\int \frac{1}{\sqrt{z - z^2}}\, dz = 4\int \frac{1}{\sqrt{1 - (2z - 1)^2}}\, dz = 2\int \frac{1}{\sqrt{1 - u^2}}\, du$.

But we have $2\int \frac{1}{\sqrt{1 - u^2}}\, du = 2\sin^{-1} u$. Or in terms of the original variable x we get $2\sin^{-1} u = 2\sin^{-1}(2z - 1) = 2\sin^{-1}\left(2\sqrt{x} - 1\right)$.

INTEGRAL 78

Problem:

$$\int x^3 \left(\ln\left(x^{\ln x - 6}\right) + 5 \right) dx$$

Solution:

$$\frac{x^4}{32}\left(8\ln^2 x - 52\ln x + 53\right) + constant$$

Techniques used:

Change of variables, Integration by parts, Trigonometric identities

Step-by-step solution:

Write the integral as a summation of its terms, $\int x^3 \left(\ln\left(x^{\ln x - 6}\right) + 5 \right)$ $dx = 5\int x^3 dx + \int x^3 \ln\left(x^{\ln x - 6}\right)dx$. But $5\int x^3 dx = \frac{5}{4}x^4$. For the remaining integral, rewrite the integrand by using logarithm rule as, $\int x^3 \ln\left(x^{\ln x - 6}\right)dx = \int x^3 (\ln x - 6)\ln x\, dx = \int x^3 \ln^2 x\, dx - 6\int x^3 \ln x\, dx$. Now, let $\ln x = z \Rightarrow x = e^z$, $dx = e^z dz$, and rewrite the new integrals in terms of the variable z. Therefore, $\int x^3 \ln^2 x\, dx = \int e^{4z} z^2 dz$. Using the integration by parts technique twice, we get $\int e^{4z} \underset{dg}{z^2}\, \underset{f}{dz} = \frac{e^{4z}}{4}z^2 - \frac{1}{2}\int e^{4z} z\, dz = \frac{e^{4z}}{4}z^2 - \frac{e^{4z}}{8}z + \frac{e^{4z}}{32}$. Similarly,

$-6\int x^3 \ln x\, dx = -6\int \underset{dg}{e^{4z}}\, \underset{f}{z}\, dz = -6\left(\frac{e^{4z}}{4}z - \frac{1}{4}\int e^{4z}\, dz\right) = -3\frac{e^{4z}}{2}z + 3\frac{e^{4z}}{8}$.

Collecting all answers and rewrite them in terms of the original variable x, we get $\frac{5}{4}x^4 + \frac{e^{4z}}{4}z^2 - \frac{e^{4z}}{8}z + \frac{e^{4z}}{32} - 3\frac{e^{4z}}{2}z + 3\frac{e^{4z}}{8} =$

$\frac{5}{4}x^4 + \frac{e^{4z}}{32}\left(8z^2 - 52z + 13\right) = \frac{5}{4}x^4 + \frac{x^4}{32}\left(8\ln^2 x - 52\ln x + 13\right)$.

The answer simplifies to $\frac{x^4}{32}\left(53 + 8\ln^2 x - 52\ln x\right)$.

INTEGRAL 79

Problem:

$$\int \tan^{-1}\left(\sqrt{x+1}-\sqrt{x}\right)dx$$

Solution:

$$\frac{1}{2}\left(\sqrt{x}-\tan^{-1}\sqrt{x}\right)+x\tan^{-1}\left(\sqrt{x+1}-\sqrt{x}\right)+constant$$

Techniques used:

Change of variables, Integration by parts, Trigonometric identities, Partial fractions

Step-by-step solution:

Using the integration by parts technique, we can write the integral as $\int \underbrace{\tan^{-1}\left(\sqrt{x+1}-\sqrt{x}\right)}_{f}\underbrace{dx}_{dg}=x\tan^{-1}\left(\sqrt{x+1}-\sqrt{x}\right)-\int x\left[\frac{d}{dx}\right.$

$\left.\tan^{-1}\left(\sqrt{x+1}-\sqrt{x}\right)\right]dx$. But having $\tan^{-1}\left(\sqrt{x+1}-\sqrt{x}\right)=\alpha$, or

$\tan\alpha=\sqrt{x+1}-\sqrt{x}$, we get, after differentiating both side, $\dfrac{d\alpha}{\cos^{2}\alpha}=\left(\dfrac{1}{2\sqrt{x+1}}-\dfrac{1}{2\sqrt{x}}\right)dx$. But we have $\tan^{2}\alpha=\dfrac{1-\cos^{2}\alpha}{\cos^{2}\alpha}=$

$\left(\sqrt{x+1}-\sqrt{x}\right)^{2}$. Or $\cos^{2}\alpha=\dfrac{1}{\left(\sqrt{x+1}-\sqrt{x}\right)^{2}+1}$. Therefore,

$\dfrac{d\alpha}{dx}=\left(\dfrac{1}{2\sqrt{x+1}}-\dfrac{1}{2\sqrt{x}}\right)\dfrac{1}{\left(\sqrt{x+1}-\sqrt{x}\right)^{2}+1}$. Now by plugging back

into the new integral, we get $-\int x\left[\dfrac{d}{dx}\tan^{-1}\left(\sqrt{x+1}-\sqrt{x}\right)\right]dx=$

$\int\left(\dfrac{1}{2\sqrt{x}}-\dfrac{1}{2\sqrt{x+1}}\right)\dfrac{x}{\left(\sqrt{x+1}-\sqrt{x}\right)^{2}+1}dx$. Now, let $\sqrt{x}=z\Rightarrow$

$dx = 2zdz$ and write the integral in terms of the variable z as

$$\int\left(\frac{1}{2\sqrt{x}}-\frac{1}{2\sqrt{x+1}}\right)\frac{x}{\left(\sqrt{x+1}-\sqrt{x}\right)^2+1}dx=\int\left(\frac{1}{z}-\frac{1}{\sqrt{z^2+1}}\right)$$

$$\frac{z^3}{\left(\sqrt{z^2+1}-z\right)^2+1}dz.$$ For performing the latter inte-

gral, let $z=\tan u \Rightarrow dz = \dfrac{du}{\cos^2 u}$. Therefore, $\displaystyle\int\left(\frac{1}{z}-\frac{1}{\sqrt{z^2+1}}\right)$

$$\frac{z^3}{\left(\sqrt{z^2+1}-z\right)^2+1}dz-\int\left(\frac{\cos u}{\sin u}-\cos u\right)\frac{\sin^3 u/\cos^3 u}{\left(1/\cos u-\sin u/\cos u\right)^2+1},$$

$$\frac{du}{\cos^2 u}=\int\frac{\sin^2 u}{\cos^2 u\left(1-\sin u\right)}du.$$ Using trigonometric half-angle

identities we can write the integral as $\displaystyle\int\frac{\sin^2 u}{\cos^2 u\left(1-\sin u\right)}du=$

$$\int\frac{4\tan^2 u/2}{\left(1-\tan^2 u/2\right)^2}\cdot\frac{1}{1-\dfrac{2\tan\dfrac{u}{2}}{1+\tan^2\dfrac{u}{2}}}du.$$ For performing the new

integral let $\tan\dfrac{u}{2}=y \Rightarrow du = \dfrac{2}{1+y^2}dy$, and rewrite the integral

in terms of the variable y as $\displaystyle\int\frac{4\tan^2 u/2}{\left(1-\tan^2 u/2\right)^2}\cdot\frac{1}{1-\dfrac{2\tan\dfrac{u}{2}}{1+\tan^2\dfrac{u}{2}}}du=$

$$8\int\frac{y^2\left(1+y^2\right)}{\left(1-y^2\right)^2\left(1+y^2-2y\right)}\cdot\frac{dy}{1+y^2}=8\int\frac{y^2}{\left(1-y\right)^4\left(1+y\right)^2}dy.$$ Now by

using the partial fractions technique, we can write the integrand as

$$\frac{y^2}{\left(1-y\right)^4\left(1+y\right)^2}=\frac{1}{16\left(y+1\right)^2}-\frac{1}{16\left(y-1\right)^2}+\frac{1}{4\left(y-1\right)^3}+\frac{1}{4\left(y-1\right)^4}$$

and the integral as a combination of its terms $8\displaystyle\int\frac{y^2}{\left(1-y\right)^4\left(1+y\right)^2}$

$$dy = \frac{1}{2}\int \frac{1}{(y+1)^2} dy - \frac{1}{2}\int \frac{1}{(y-1)^2} dy + 2\int \frac{1}{(y-1)^3} dy + 2\int \frac{1}{(y-1)^4} dy.$$

The new integrals can be worked out as $\dfrac{1}{2}\int \dfrac{1}{(y+1)^2} dy = -\dfrac{1}{2(y+1)}$,

$$-\frac{1}{2}\int \frac{1}{(y-1)^2} dy = \frac{1}{2(y-1)}, \qquad 2\int \frac{1}{(y-1)^3} dy = -\frac{1}{(y-1)^2}, \qquad \text{and}$$

$$2\int \frac{1}{(y-1)^4} dy = -\frac{2}{3(y-1)^3}.$$ by subsequent change of variables

$\left(y = \tan \dfrac{u}{2}, z = \tan u = \sqrt{x}\right)$ we get $y = \dfrac{\sqrt{1+x} - 1}{\sqrt{x}}$, assuming $y > 0$.

Therefore, after collecting all results and writing them in terms of the original variable x, we get the answer as

$$-x \tan^{-1}\left(\sqrt{x+1} - \sqrt{x}\right) + \frac{1}{2}\left(\sqrt{x} - \tan^{-1}\sqrt{x}\right).$$

INTEGRAL 80

Problem:

$$\int \frac{x^9}{x^{20} - 48x^{10} + 575} dx$$

Solution:

$$\frac{1}{20} \ln\left(\frac{x^{10} - 25}{x^{10} - 23} \right) + constant$$

Techniques used:

Change of variables, Integration by parts, Trigonometric identities

Step-by-step solution:

The denominator of the integrand can be written as $x^{20} - 48x^{10} + 575 = \left(x^{10} - 25\right)\left(x^{10} - 23\right)$. Therefore, using the partial fractions technique we get $\int \frac{x^9}{x^{20} - 48x^{10} + 575} dx = \frac{1}{2}\int \frac{x^9}{x^{10} - 25} dx$ $- \frac{1}{2}\int \frac{x^9}{x^{10} - 23} dx$. Or $\frac{1}{2}\int \frac{x^9}{x^{10} - 25} dx = \frac{1}{20}\ln\left(x^{10} - 25\right)$ and $-\frac{1}{2}$ $\int \frac{x^9}{x^{10} - 23} dx = -\frac{1}{20}\ln\left(x^{10} - 23\right)$. Collecting the answers, we get the solution as $\frac{1}{20}\ln\frac{x^{10} - 25}{x^{10} - 23}$.

INTEGRAL 81

Problem:

$$\int \sin^{-1} x \, dx$$

Solution:

$$x \sin^{-1} x + \sqrt{1 - x^2} + constant$$

Techniques used:

Change of variables, Integration by parts, Trigonometric identities

Step-by-step solution:

Let $\sin^{-1} x = \alpha \Rightarrow x = \sin \alpha$ and $dx = \cos \alpha \, d\alpha$. Therefore, we can write the integral in terms of the variable α as $\int \sin^{-1} x \, dx = \int \alpha \cos \alpha \, d\alpha$. Using the integration by parts technique, we get $\int \underset{f}{\alpha} \underset{dg}{\cos \alpha} \, d\alpha = \alpha \sin \alpha - \int \sin \alpha \, d\alpha = \alpha \sin \alpha + \cos \alpha$. The results in terms of the original variable x reads, $\alpha \sin \alpha + \cos \alpha = x \sin^{-1} x + \sqrt{1 - x^2}$.

INTEGRAL 82

Problem:

$$\int \tan^{-1} x \, dx$$

Solution:

$$x \tan^{-1} x - \frac{1}{2} \ln\left(1 + x^2\right) + constant$$

Techniques used:

Change of variables, Integration by parts, Trigonometric identities

Step-by-step solution:

Let $\tan^{-1} x = \alpha \Rightarrow x = \tan\alpha$ and $dx = \cos^{-2}\alpha \, d\alpha$. Therefore, we can write the integral in terms of variable α as $\int \tan^{-1} x \, dx = \int \alpha \cos^{-2}\alpha \, d\alpha$. Using the integration by parts technique, we get $\int \underset{f}{\underline{\alpha}} \underset{dg}{\underline{\cos^{-2}\alpha}} \, d\alpha = \alpha \tan\alpha - \int \tan\alpha \, d\alpha$. But $-\int \tan\alpha \, d\alpha = -\int \frac{\sin\alpha}{\cos\alpha} \, d\alpha$

$= \ln\cos\alpha$. The results in terms of the original variable x reads,

$$\alpha \tan\alpha + \ln\cos\alpha = x \tan^{-1} x + \ln\left(\frac{1}{\sqrt{1 + x^2}}\right) = x \tan^{-1} x - \frac{1}{2} \ln\left(1 + x^2\right).$$

Note that having $x = \tan\alpha$ we can calculate, $\cos\alpha = \frac{1}{\sqrt{1 + x^2}}$.

INTEGRAL 83

Problem:

$$\int \sinh^{-1} x \, dx$$

Solution:

$$x \sinh^{-1} x - \sqrt{1 + x^2} + constant$$

Techniques used:

Change of variables, Integration by parts, Trigonometric identities

Step-by-step solution:

Let $\sinh^{-1} x = \alpha \Rightarrow x = \sinh \alpha$ and $dx = \cosh \alpha \, d\alpha$. Therefore, we can write the integral in terms of variable α as $\int \sinh^{-1} x \, dx = \int \alpha \cosh \alpha \, d\alpha$. Using the integration by parts technique, we get $\int \underset{\widetilde{f}}{\alpha} \underset{\widetilde{dg}}{\cosh \alpha \, d\alpha} = \alpha \sinh \alpha - \int \sinh \alpha \, d\alpha = \alpha \sinh \alpha - \cosh \alpha$. The results in terms of the original variable x reads, $\alpha \sinh \alpha - \cosh \alpha = x \sinh^{-1} x - \sqrt{1 + x^2}$. Note that $\cosh \alpha = \sqrt{1 + x^2}$, using the trigonometric identity $\cosh^2 \alpha - \underset{x^2}{\underbrace{\sinh^2 \alpha}} = 1$.

INTEGRAL 84

Problem:

$$\int \tanh^{-1} x \, dx$$

Solution:

$$x \tanh^{-1} x + \frac{1}{2} \ln\left(1 - x^2\right) + constant$$

Techniques used:

Change of variables, Integration by parts, Trigonometric identities

Step-by-step solution:

Let $\tanh^{-1} x = \alpha \Rightarrow x = \tanh\alpha$ and $dx = \cosh^{-2}\alpha \, d\alpha$. Therefore, we can write the integral in terms of variable α as $\int \tanh^{-1} x \, dx = \int \alpha \cosh^{-2}\alpha \, d\alpha$. Using the integration by parts technique, we get $\int \underbrace{\alpha}_{f} \underbrace{\cosh^{-2}\alpha}_{dg} \, d\alpha = \alpha \tanh\alpha - \int \tanh\alpha \, d\alpha$. But

$-\int \tanh\alpha \, d\alpha = -\int \dfrac{\sinh\alpha}{\cosh\alpha} d\alpha = -\ln\cosh\alpha$. The results in terms of the original variable x reads, $\alpha \tanh\alpha - \ln\cosh\alpha = x \tanh^{-1} x -$

$\ln\left(\dfrac{1}{\sqrt{1-x^2}}\right) = x \tanh^{-1} x + \dfrac{1}{2}\ln\left(1 - x^2\right)$. Note that having $x = \tanh\alpha$

we can calculate, $\cosh\alpha = \dfrac{1}{\sqrt{1-x^2}}$.

INTEGRAL 85

Problem:

$$\int \cos^{-1}\left(\frac{1}{x}\right) dx$$

Solution:

$$x\cos^{-1}\left(\frac{1}{x}\right) - \tanh^{-1}\left(\frac{\sqrt{x^2-1}}{x}\right) + constant$$

Techniques used:

Change of variables, Integration by parts, Trigonometric identities

Step-by-step solution:

Let $\cos^{-1}(1/x) = \alpha \Rightarrow x = 1/\cos\alpha$ and $dx = \sin\alpha / \cos^2\alpha\, d\alpha$. Therefore, we can write the integral in terms of the variable α as $\int \cos^{-1}(1/x) dx = \int \alpha\left(\sin\alpha / \cos^2\alpha\right) d\alpha$. Using the integration by parts technique, we get $\int \underbrace{\alpha}_{f} \underbrace{\left(\sin\alpha / \cos^2\alpha\right) d\alpha}_{dg} = \alpha / \cos\alpha - \int \frac{1}{\cos\alpha}\, d\alpha.$

But $\int \dfrac{1}{\cos\alpha} d\alpha = \tanh^{-1}(\sin\alpha)$ (see Integral 5). Therefore, we can write the results as $\alpha / \cos\alpha - \int \dfrac{1}{\cos\alpha} d\alpha = \alpha / \cos\alpha - \tanh^{-1}(\sin\alpha).$

The results in terms of the original variable x reads,

$$\alpha / \cos\alpha - \tanh^{-1}(\sin\alpha) = x\cos^{-1}(1/x) - \tanh^{-1}\left(\frac{\sqrt{x^2-1}}{|x|}\right).$$

INTEGRAL 86

Problem:

$$\int \frac{1}{x^4 + 4} dx$$

Solution:

$$\frac{1}{16} \ln \left(\frac{x^2 + 2x + 2}{x^2 - 2x + 2} \right) + \frac{1}{8} \left[\tan^{-1}(x+1) + \tan^{-1}(x-1) \right] + constant$$

Techniques used:

Change of variables, Integration by parts, Partial fractions

Step-by-step solution:

Write the denominator of the integrand as $x^4 + 4 = \left(x^2 + 2 \right)^2 - 4x^2 =$ $\left(x^2 + 2x + 2 \right) \left(x^2 - 2x + 2 \right)$. Using the partial fractions techniques, write the integral as $\int \frac{1}{x^4 + 4} dx = \frac{1}{8} \int \frac{2 + x}{x^2 + 2x + 2} dx + \frac{1}{8} \int \frac{2 - x}{x^2 - 2x + 2} dx$.

But $\frac{1}{8} \int \frac{2 + x}{x^2 + 2x + 2} dx = \frac{1}{8} \int \frac{1 + 1 + x}{x^2 + 2x + 2} dx = \frac{1}{8} \int \frac{1}{x^2 + 2x + 2} dx + \frac{1}{16}$ $\int \frac{2 + 2x}{x^2 + 2x + 2} dx$. The latter integral can be worked out, since $\frac{d}{dx} \left(x^2 + 2x + 2 \right) = 2x + 2$, as $\frac{1}{16} \int \frac{2 + 2x}{x^2 + 2x + 2} dx = \frac{1}{16} \ln \left(x^2 + 2x + 2 \right)$.

For the former integral, write it as $\frac{1}{8} \int \frac{1}{x^2 + 2x + 2} dx =$ $\frac{1}{8} \int \frac{1}{(x+1)^2 + 1} dx$. Now let $x + 1 = z \Rightarrow dx = dz$ and write the integral in terms of variable z as $\frac{1}{8} \int \frac{1}{(x+1)^2 + 1} dx = \frac{1}{8} \int \frac{1}{z^2 + 1} dz =$

$: \dfrac{1}{8}\tan^{-1}z = \dfrac{1}{8}\tan^{-1}(x+1)$. Similarly, The remaining integral

can be worked out as $\dfrac{1}{8}\displaystyle\int \dfrac{2-x}{x^2-2x+2}dx = -\dfrac{1}{8}\displaystyle\int \dfrac{x-1-1}{x^2-2x+2}dx = -$

$\cdot\dfrac{1}{16}\displaystyle\int \dfrac{2x-2}{x^2-2x+2}dx + \dfrac{1}{8}\displaystyle\int \dfrac{1}{x^2-2x+2}dx.$ But $-\dfrac{1}{16}\displaystyle\int \dfrac{2x-2}{x^2-2x+2}dx = -$

$\dfrac{1}{16}\ln\left(x^2-2x+2\right).$ For the former integral $\dfrac{1}{8}\displaystyle\int \dfrac{1}{x^2-2x+2}dx =$

$\dfrac{1}{8}\displaystyle\int \dfrac{1}{(x-1)^2+1}dx = \dfrac{1}{8}\tan^{-1}(x-1).$ Collecting all related terms,

we receive the answer as $\dfrac{1}{16}\ln\left(x^2+2x+2\right) + \dfrac{1}{8}\tan^{-1}(x+1) -$

$\dfrac{1}{16}\ln\left(x^2-2x+2\right) + \dfrac{1}{8}\tan^{-1}(x-1).$ This expression simplifies to

$\dfrac{1}{16}\ln\left(\dfrac{x^2+2x+2}{x^2-2x+2}\right) + \dfrac{1}{8}\left[\tan^{-1}(x+1) + \tan^{-1}(x-1)\right].$

INTEGRAL 87

Problem:

$$\int \frac{x^5}{\left(4x^2+4\right)^{5/2}}\,dx$$

Solution:

$$\frac{1}{12}\left[\frac{3x^4+12x^2+8}{\left(4x^2+4\right)^{3/2}}\right]+constant$$

Techniques used:

Change of variables, Integration by parts

Step-by-step solution:

Expand the denominator of the integrand as $\left(4x^2+4\right)^{5/2}=$ $4^{5/2}\left(x^2+1\right)^{5/2}=32\sqrt{x^2+1}\left(x^2+1\right)^2$. Therefore, the integral reads $\int \frac{x^5}{\left(4x^2+4\right)^{5/2}}\,dx=\frac{1}{32}\int \frac{x^5}{\sqrt{x^2+1}\left(x^2+1\right)^2}\,dx$. Now, using the partial fractions technique[1], we get $\frac{1}{32}\int \frac{x^5}{\sqrt{x^2+1}\left(x^2+1\right)^2}\,dx=$

$$\frac{1}{32}\left[\int \frac{x}{\sqrt{x^2+1}}\,dx-\int \frac{2x}{\left(x^2+1\right)^{3/2}}\,dx+\int \frac{x}{\left(x^2+1\right)^{5/2}}\,dx\right].\text{Now,perform}$$

the new integrals, one by one. For $\frac{1}{32}\int \frac{x}{\sqrt{x^2+1}}\,dx=$

$\frac{1}{64}\int \frac{2x}{\sqrt{x^2+1}}\,dx=\frac{\sqrt{x^2+1}}{32}$. For $-\frac{1}{32}\int \frac{2x}{\left(x^2+1\right)^{3/2}}\,dx=\frac{1}{16\sqrt{x^2+1}}$. For

[1] A CAS tool (e.g., Wolfram Alpha) can be employed for this step.

$$\frac{1}{32}\int \frac{x}{\left(x^2+1\right)^{5/2}}\,dx = \frac{1}{64}\int \frac{2x}{\left(x^2+1\right)^{5/2}}\,dx = -\frac{1}{96\left(x^2+1\right)^{3/2}}.$$ Collecting all related results, we receive the answer as $\frac{\sqrt{x^2+1}}{32}+\frac{1}{16\sqrt{x^2+1}}-\frac{1}{96\left(x^2+1\right)^{3/2}}.$ This expression simplifies to

$$\frac{3x^4+12x^2+8}{96\left(x^2+1\right)\sqrt{x^2+1}} = \frac{1}{12}\left[\frac{3x^4+12x^2+8}{\left(4x^2+4\right)^{3/2}}\right].$$

INTEGRAL 88

Problem:

$$\int \frac{1}{x^4 + 1} dx$$

Solution:

$$\frac{\sqrt{2}}{8} \left\{ \ln\left(\frac{x^2 + \sqrt{2}x + 1}{x^2 - \sqrt{2}x + 1}\right) + 2\left[\tan^{-1}\left(\sqrt{2}x + 1\right) + \right.\right.$$

$$\left.\left. \tan^{-1}\left(\sqrt{2}x - 1\right)\right]\right\} + constant$$

Techniques used:

Change of variables, Integration by parts, Partial fractions

Step-by-step solution:

Write the expression in the denominator of the integral as a product of two terms. Therefore, $x^4 + 1 = \left(x^2 + 1\right)^2 - 2x^2 = \left(x^2 + \sqrt{2}x + 1\right)$ $\left(x^2 - \sqrt{2}x + 1\right)$. But, using the partial fractions technique, we can write the integral as $\int \frac{1}{x^4 + 1} dx = \int \frac{1}{\left(x^2 + \sqrt{2}x + 1\right)\left(x^2 - \sqrt{2}x + 1\right)} dx$

$= \frac{\sqrt{2}}{4} \int \frac{x + \sqrt{2}}{x^2 + \sqrt{2}x + 1} dx - \frac{\sqrt{2}}{4} \int \frac{x - \sqrt{2}}{x^2 - \sqrt{2}x + 1} dx$. Now we perform the

integration of the new integrals. Rewrite the former integral as

$\frac{\sqrt{2}}{4} \int \frac{x + \sqrt{2}}{x^2 + \sqrt{2}x + 1} dx = \frac{\sqrt{2}}{8} \int \frac{2x + \sqrt{2} + \sqrt{2}}{x^2 + \sqrt{2}x + 1} dx = \frac{\sqrt{2}}{8} \left(\int \frac{2x + \sqrt{2}}{x^2 + \sqrt{2}x + 1} \right.$

$\left. dx + \int \frac{\sqrt{2}}{x^2 + \sqrt{2}x + 1} dx \right)$. Therefore, $\frac{\sqrt{2}}{8} \int \frac{2x + \sqrt{2}}{x^2 + \sqrt{2}x + 1} dx = \frac{\sqrt{2}}{8} \ln$

$\left(x^2 + \sqrt{2}x + 1\right)$. Note that $\dfrac{d}{dx}\left(x^2 + \sqrt{2}x + 1\right) = 2x + \sqrt{2}$. Now, after manipulating the expression in the denominator of the latter integral, we have $x^2 + \sqrt{2}x + 1 = \left(x + \dfrac{\sqrt{2}}{2}\right)^2 + \dfrac{1}{2} = \dfrac{1}{2}\left[1 + \left(\sqrt{2}x + 1\right)^2\right]$.

Therefore, we have $\dfrac{\sqrt{2}}{8}\displaystyle\int \dfrac{\sqrt{2}}{x^2 + \sqrt{2}x + 1}\,dx = \dfrac{1}{2}\displaystyle\int \dfrac{1}{1 + \left(\sqrt{2}x + 1\right)^2}\,dx$.

Let $\sqrt{2}x + 1 = z \Rightarrow \sqrt{2}\,dx = dz$ and write the integral in terms of the variable z, as $\dfrac{1}{2}\displaystyle\int \dfrac{1}{1 + \left(\sqrt{2}x + 1\right)^2}\,dx = \dfrac{1}{2\sqrt{2}}\displaystyle\int \dfrac{1}{1 + z^2}\,dz = \dfrac{\sqrt{2}}{4}\tan^{-1} z$.

Or in terms of the original variable, x we get $\dfrac{\sqrt{2}}{4}\tan^{-1}\left(\sqrt{2}x + 1\right)$.

Similarly, for the remaining integral, we have

$-\dfrac{\sqrt{2}}{4}\displaystyle\int \dfrac{x - \sqrt{2}}{x^2 - \sqrt{2}x + 1}\,dx = -\dfrac{\sqrt{2}}{8}\displaystyle\int \dfrac{2x - \sqrt{2} - \sqrt{2}}{x^2 - \sqrt{2}x + 1}\,dx = \dfrac{\sqrt{2}}{8}$

$\left(-\displaystyle\int \dfrac{2x - \sqrt{2}}{x^2 - \sqrt{2}x + 1}\,dx + \displaystyle\int \dfrac{\sqrt{2}}{x^2 - \sqrt{2}x + 1}\,dx\right)$.

Therefore, $-\dfrac{\sqrt{2}}{8}\displaystyle\int \dfrac{2x - \sqrt{2}}{x^2 - \sqrt{2}x + 1}\,dx = -\dfrac{\sqrt{2}}{8}\ln\left(x^2 - \sqrt{2}x + 1\right)$. Note

that $\dfrac{d}{dx}\left(x^2 - \sqrt{2}x + 1\right) = 2x - \sqrt{2}$. Now, after manipulating the expression in the denominator of the new integral, we have

$x^2 - \sqrt{2}x + 1 = \left(x - \dfrac{\sqrt{2}}{2}\right)^2 + \dfrac{1}{2} = \dfrac{1}{2}\left[1 + \left(\sqrt{2}x - 1\right)^2\right]$. Therefore, we

have $\dfrac{\sqrt{2}}{8}\displaystyle\int \dfrac{\sqrt{2}}{x^2 - \sqrt{2}x + 1}\,dx = \dfrac{1}{2}\displaystyle\int \dfrac{1}{1 + \left(\sqrt{2}x - 1\right)^2}\,dx$. Let $\sqrt{2}x - 1 = u \Rightarrow$

$\sqrt{2}\,dx = du$ and write the integral in terms of the variable u, as

$\dfrac{1}{2}\displaystyle\int \dfrac{1}{1 + \left(\sqrt{2}x - 1\right)^2}\,dx = \dfrac{1}{2\sqrt{2}}\displaystyle\int \dfrac{1}{1 + u^2}\,du = \dfrac{\sqrt{2}}{4}\tan^{-1} u$. Or in terms of the original variable x, we get $\dfrac{\sqrt{2}}{4}\tan^{-1}\left(\sqrt{2}x-1\right)$. Collecting all related results, we get the answer as

$$\frac{\sqrt{2}}{8}\ln\left(x^2+\sqrt{2}x+1\right)+\frac{\sqrt{2}}{4}\tan^{-1}\left(\sqrt{2}x+1\right)-\frac{\sqrt{2}}{8}\ln\left(x^2-\sqrt{2}x+1\right)+$$

$$\frac{\sqrt{2}}{4}\tan^{-1}\left(\sqrt{2}x-1\right). \text{ This expression simplifies to}$$

$$\frac{\sqrt{2}}{8}\left\{\ln\frac{\left(x^2+\sqrt{2}x+1\right)}{\left(x^2-\sqrt{2}x+1\right)}+2\left[\tan^{-1}\left(\sqrt{2}x+1\right)+\tan^{-1}\left(\sqrt{2}x-1\right)\right]\right\}.$$

INTEGRAL 89

Problem:

$$\int \sin^2 (\ln x) dx$$

Solution:

$$\frac{x}{10}\left[5 - \cos(2\ln x) - 2\sin(2\ln x)\right] + constant$$

Techniques used:

Change of variables, Integration by parts, Trigonometric identities

Step-by-step solution:

Let $\ln x = z \Rightarrow dx = e^z dz$. Now write the integral in terms of variable z, as $\int \sin^2 (\ln x) dx = \int e^z \sin^2 z \, dz$. Using the trigonometric identity, $\sin^2 z = \dfrac{1 - \cos 2z}{2}$ rewrite the integral as $\dfrac{1}{2}\int e^z (1 - \cos 2z) dz = \dfrac{1}{2}\int e^z dz - \dfrac{1}{2}\int e^z \cos 2z \, dz$. For the new integrals, we get $\dfrac{1}{2}\int e^z dz = \dfrac{1}{2}e^z$ and, using the integration by parts technique,

$$-\frac{1}{2}\underbrace{\int e^z}_{f} \underbrace{\cos 2z}_{dg} dz = -\frac{1}{4}e^z \sin 2z + \frac{1}{4}\int e^z \sin 2z \, dz = -\frac{1}{4}e^z \sin 2z +$$

$-\dfrac{1}{8}e^z \cos 2z + \dfrac{1}{8}\int e^z \cos 2z \, dz$. Therefore, rearranging terms in the last expression gives $\int e^z \cos 2z \, dz = \dfrac{1}{5}e^z (\cos 2z + 2\sin 2z)$. Now, we can write $\int e^z \sin^2 z \, dz = \dfrac{1}{2}e^z - \dfrac{1}{10}e^z (\cos 2z + 2\sin 2z)$. This expression written in the original variable x gives the answer as $\dfrac{1}{2}e^z - \dfrac{1}{10}e^z (\cos 2z + 2\sin 2z) = x\left(\dfrac{1}{2} - \dfrac{1}{10}(\cos(2\ln x) + 2\sin(2\ln x))\right)$.

INTEGRAL 90

Problem:

$$\int \sin x \sin 2x \sin 3x \, dx$$

Solution:

$$\frac{1}{24}\left(6\sin^2 x + 3\sin^2 2x - 2\sin^2 3x\right) + constant$$

Or

$$-\frac{1}{24}\left(6\cos^2 x + 3\cos^2 2x - \cos^2 3x\right) + constant$$

Techniques used:

Integration by parts, Trigonometric identities

Step-by-step solution:

Using the trigonometric identity, $\sin a \sin b = \frac{1}{2}\left[\cos(a-b) - \cos(a+b)\right]$, rewrite the integrand as $(\sin x \sin 2x)\sin 3x = \frac{1}{2}[\cos x - \underbrace{}_{b}\underbrace{}_{a}$

$\cos 3x]\sin 3x = \frac{1}{2}[\cos x \sin 3x - \cos 3x \sin 3x]$. Now, after substitution, rewrite the integral as $\int \sin x \sin 2x \sin 3x \, dx = \frac{1}{2}\int \cos x \sin 3x \, dx - \frac{1}{2}\int \cos 3x \sin 3x \, dx$. For the latter new integral, using the integration by parts, we get $-\frac{1}{2}\int \cos 3x \sin 3x dx = \frac{1}{12}\cos^2 3x$. For the former integral, using the trigonometric identity, $\sin c \cos d = \frac{1}{2}\left[\sin(c+d) + \sin(c-d)\right]$, we have $\frac{1}{2}\int \underbrace{\cos x}_{d} \underbrace{\sin 3x}_{c} \, dx =$

$$\frac{1}{4}\int\left(\sin 4x + \sin 2x\right)dx = \frac{1}{2}\int \sin 2x \cos 2x\, dx + \frac{1}{2}\int \sin x \cos x\, dx.$$

The new integrals are worked out as $\dfrac{1}{2}\int \sin 2x \cos 2x\, dx = \dfrac{1}{8}\sin^2 2x,$

and $\dfrac{1}{2}\int \sin x \cos x\, dx = \dfrac{1}{4}\sin^2 x.$ Collecting all related terms, the

answer reads $\dfrac{1}{12}\cos^2 3x + \dfrac{1}{8}\sin^2 2x + \dfrac{1}{4}\sin^2 x.$ This expression simpli-

fies to $\dfrac{1}{12} + \dfrac{1}{24}\left(6\sin^2 x + 3\sin^2 2x - 2\sin^2 3x\right).$ Or the answer in terms

of the cosine function reads $\dfrac{3}{8} - \dfrac{1}{24}\left(6\cos^2 x + 3\cos^2 2x - \cos^2 3x\right).$

INTEGRAL 91

Problem:

$$\int \sqrt{1+x^2}\,dx$$

Solution:

$$\frac{1}{2}\left(\sinh^{-1}x + x\sqrt{1+x^2}\right) + constant$$

Techniques used:

Change of variables, Trigonometric identities

Step-by-step solution:

Let $x = \sinh z \Rightarrow dx = \cosh z\,dz$, and write the integral in terms of the variable z as $\int \sqrt{1+x^2}\,dx = \int \sqrt{1+\sinh^2 z}\,\cosh z\,dz = \int \cosh^2 z\,dz$. Note that $\cosh^2 z - \sinh^2 z = 1$. Rewriting the latter integral using the trigonometric identity, $\cosh^2 z = \frac{1}{2}(\cosh 2z + 1)$, as $\int \cosh^2 z\,dz = \frac{1}{2}\int (\cosh 2z + 1)\,dz = \frac{1}{2}\int \cosh 2z\,dz + \frac{z}{2}$. But $\frac{1}{2}\int \cosh 2z\,dz = \frac{1}{4}\sinh 2z$. Collecting related terms, the results read $\frac{z}{2} + \frac{1}{4}\sinh 2z$. This expression in terms of the original variable x reads $\frac{\sinh^{-1} x}{2} + \frac{1}{2}\sinh z\cosh z = \frac{1}{2}\left(\sinh^{-1} x + x\sqrt{1+x^2}\right)$.

INTEGRAL 92

Problem:

$$\int \frac{\ln x \cos x - \frac{1}{x}\sin x}{\left(\ln x\right)^2}\,dx$$

Solution:

$$\frac{\sin x}{\ln x} + constant$$

Techniques used:

Integration by parts

Step-by-step solution:

Expand the integrand, $\dfrac{\ln x \cos x - \frac{1}{x}\sin x}{\left(\ln x\right)^2} = \dfrac{\cos x}{\ln x} - \dfrac{\sin x}{x \ln^2 x}$. Therefore,

$\displaystyle\int \frac{\ln x \cos x - \frac{1}{x}\sin x}{\left(\ln x\right)^2}\,dx = \int \frac{\cos x}{\ln x}\,dx - \int \frac{\sin x}{x \ln^2 x}\,dx$. Integrate former

integral by using the integration by parts technique, to get

$\displaystyle\int \underset{dg}{\underbrace{\cos x}}\left(\underset{f}{\underbrace{\frac{1}{\ln x}}}\right)dx = \frac{\sin x}{\ln x} + \int \frac{\sin x}{x \ln^2 x}\,dx$. Collecting all related terms

gives $\dfrac{\sin x}{\ln x} + \underset{=0}{\underbrace{\int \frac{\sin x}{x \ln^2 x}\,dx - \int \frac{\sin x}{x \ln^2 x}\,dx}} = \dfrac{\sin x}{\ln x}$.

INTEGRAL 93

Problem:

$$\int \frac{2e^{2x} - e^x}{\sqrt{3e^{2x} - 6e^x - 1}} dx$$

Solution:

$$\frac{1}{3}\left[2\sqrt{3e^{2x} - 6e^x - 1} + \sqrt{3}\cosh^{-1}\left(\frac{\sqrt{3}e^x - \sqrt{3}}{2} \right) \right] + constant$$

Techniques used:

Integration by parts, Trigonometric identities

Step-by-step solution:

Let $e^x = z \Rightarrow dx = dz / z$ and rewrite the integral in terms of the variable z, to get $\int \frac{2e^{2x} - e^x}{\sqrt{3e^{2x} - 6e^x - 1}} dx = \int \frac{2z^2 - z}{z\sqrt{3z^2 - 6z - 1}} dz =$

$\int \frac{2z - 1}{\sqrt{3z^2 - 6z - 1}} dz$. The expression in the denominator can be writ-

ten as $\sqrt{3z^2 - 6z - 1} = \sqrt{3}\left(\sqrt{z^2 - 2z - 1/3} \right) = \sqrt{3}\left(\sqrt{(z-1)^2 - 4/3} \right)$

$= 2\sqrt{\left[\frac{\sqrt{3}(z-1)}{2} \right]^2 - 1}$. Therefore, the integral reads

$\int \frac{2z - 1}{\sqrt{3z^2 - 6z - 1}} dz = \frac{1}{2}\int \frac{2z - 1}{\sqrt{\left[\frac{\sqrt{3}(z-1)}{2} \right]^2 - 1}} dz$. Let $\frac{\sqrt{3}(z-1)}{2} =$

$u \Rightarrow dz = \frac{2}{\sqrt{3}} du,\ z = \frac{2}{\sqrt{3}}u + 1$. Now, write the latter integral in terms of the variable u, as $\dfrac{1}{2}\displaystyle\int \dfrac{2z-1}{\sqrt{\left[\dfrac{\sqrt{3}(z-1)}{2}\right]^2 - 1}}\,dz = \dfrac{1}{3}$

$\displaystyle\int \dfrac{4u+\sqrt{3}}{\sqrt{u^2-1}}\,du.$ Expand the new integral, as $\dfrac{1}{3}\displaystyle\int \dfrac{4u+\sqrt{3}}{\sqrt{u^2-1}}\,du = \dfrac{4}{3}$

$\displaystyle\int \dfrac{u}{\sqrt{u^2-1}}\,du + \dfrac{\sqrt{3}}{3}\displaystyle\int \dfrac{1}{\sqrt{u^2-1}}\,du.$ For the former integral, we

have $\dfrac{4}{3}\displaystyle\int \dfrac{u}{\sqrt{u^2-1}}\,du = \dfrac{4}{3}\sqrt{u^2-1}.$ For the latter integral, let

$u = \cosh y \Rightarrow du = \sinh y\, dy$ and write the integral in terms of the

variable y, as $\dfrac{\sqrt{3}}{3}\displaystyle\int \dfrac{1}{\sqrt{u^2-1}}\,du = \dfrac{\sqrt{3}}{3}\displaystyle\int \dfrac{\sinh y}{\sqrt{\cosh^2 y - 1}}\,dy = \dfrac{\sqrt{3}}{3}\displaystyle\int dy =$

$\dfrac{\sqrt{3}}{3}y = \dfrac{\sqrt{3}}{3}\cosh^{-1} u.$ Collecting all related terms and write them in

terms of the original variable x, gives $\dfrac{4}{3}\sqrt{u^2-1} + \dfrac{\sqrt{3}}{3}\cosh^{-1} u = \dfrac{4}{3}$

$\sqrt{\left(\dfrac{\sqrt{3}(z-1)}{2}\right)^2 - 1} + \dfrac{\sqrt{3}}{3}\cosh^{-1}\left(\dfrac{\sqrt{3}(z-1)}{2}\right) = \dfrac{4}{3}\sqrt{\left(\dfrac{\sqrt{3}(e^x-1)}{2}\right)^2 - 1}$

$+ \dfrac{\sqrt{3}}{3}\cosh^{-1}\left(\dfrac{\sqrt{3}(e^x-1)}{2}\right).$ This expression simplifies to

$\dfrac{2}{3}\sqrt{3e^{2x}-6e^x-1} + \dfrac{1}{\sqrt{3}}\cosh^{-1}\left(\dfrac{\sqrt{3}(e^x-1)}{2}\right).$

INTEGRAL 94

Problem:

$$\int \left(\frac{x^4}{1+x^6} \right)^2 dx$$

Solution:

$$\frac{1}{6}\left(\tan^{-1} x^3 - \frac{x^3}{1+x^6} \right) + constant$$

Techniques used:

Change of variables, Integration by parts, Trigonometric identities

Step-by-step solution:

Let $x^3 = z \Rightarrow 3x^2 dx = dz$ and write the integral in terms of the variable z, as $\int \left(\frac{x^4}{1+x^6} \right)^2 dx = \int \left(\frac{x^4}{1+z^2} \right)^2 \frac{dz}{3x^2} = \frac{1}{3}\int \left(\frac{z}{1+z^2} \right)^2 dz =$

$\frac{1}{3}\int \frac{z^2}{\left(1+z^2\right)^2} dz$. Now let $z = \tan u \Rightarrow dz = \frac{du}{\cos^2 u}$ and write the new

integral in terms of the variable u, as $\frac{1}{3}\int \frac{\tan^2 u}{\cos^2 u \left(1 + \tan^2 u\right)^2} du =$

$\frac{1}{3}\int \sin^2 u\, du$. But $\sin^2 u = \frac{1-\cos 2u}{2}$. Therefore, we have $\frac{1}{3}\int \sin^2 u\, du =$

$\frac{1}{6}\int (1 - \cos 2u)\, du = \frac{u}{6} - \frac{1}{12}\sin 2u$. Therefore, the answer in

terms of the original variable x reads $\frac{u}{6} - \frac{1}{12}\sin 2u = \frac{\tan^{-1} z}{6} -$

$\frac{1}{12}\sin\left(2\tan^{-1} z\right) = \frac{\tan^{-1} x^3}{6} - \frac{1}{12}\sin\left(2\tan^{-1} x^3\right)$. But $\sin\left(2\tan^{-1} x^3\right) =$

$\frac{2x^3}{1+x^6}$, using $\tan^{-1} x^3 = \alpha$, or $\tan\alpha = x^3$, and calculating $\sin 2\alpha$. The

answer simplifies to $\frac{\tan^{-1} x^3}{6} - \frac{x^3}{6\left(1+x^6\right)}$.

INTEGRAL 95

Problem:

$$\int \frac{x^3 e^{x^2}}{\left(1+x^2\right)^2} dx$$

Solution:

$$\frac{e^{x^2}}{2\left(1+x^2\right)} + constant$$

Techniques used:

Change of variables, Integration by parts

Step-by-step solution:

Let $x^2 = z \Rightarrow 2xdx = dz$ and write the integral in terms of variable z, as $\int \frac{x^3 e^{x^2}}{\left(1+x^2\right)^2} dx = \frac{1}{2}\int \frac{ze^z}{\left(1+z\right)^2} dz$. Now, using the integration by parts technique we get $\frac{1}{2}\int \underbrace{ze^z}_{f} \underbrace{\frac{dz}{\left(1+z\right)^2}}_{dg} = -\frac{1}{2}ze^z\left(\frac{1}{1+z}\right) +$

$\frac{1}{2}\int \frac{e^z\left(1+z\right)}{1+z}dz = -\frac{1}{2}\frac{ze^z}{1+z} + \frac{1}{2}e^z = \frac{1}{2}e^z\left(\frac{1}{1+z}\right)$. This expression in terms of the original variable x gives the answer as $\frac{1}{2}e^z\left(\frac{1}{1+z}\right) = \frac{e^{x^2}}{2\left(1+x^2\right)}$.

INTEGRAL 96

Problem:

$$\int \frac{1}{1+\sqrt{\sqrt{x}}} dx$$

Solution:

$$\frac{4\sqrt[4]{x^3}}{3} - 2\sqrt{x} + 4\sqrt[4]{x} - 4\ln\left(1 + \sqrt[4]{x}\right) + constant$$

Techniques used:

Change of variables, Integration by parts

Step-by-step solution:

Let $\sqrt{x} = z \Rightarrow dx = 2zdz$ and write the integral in terms of the variable z, as $\int \frac{1}{1+\sqrt{\sqrt{x}}} dx = 2\int \frac{z}{1+\sqrt{z}} dz$. Now, let $1 + \sqrt{z} = u \Rightarrow dz = 2(u-1)du$ and the new integral in terms of the variable u reads $2\int \frac{z}{1+\sqrt{z}} dz = 4\int \frac{(u-1)^3}{u} du$. After expanding the polynomial, we have $(u-1)^3 = u^3 - 3u^2 + 3u - 1$. Therefore, $4\int \frac{(u-1)^3}{u} du =$

$$4\int \frac{u^3 - 3u^2 + 3u - 1}{u} du = 4\int u^2 du - 12\int u\, du + 12\int du - 4\int \frac{1}{u} du.$$

Performing the new integrals, gives $\frac{4}{3}u^3 - 6u^2 + 12u - 4\ln u$. Write this expression in terms of the original variable x gives the answer as

$$\frac{4}{3}u^3 - 6u^2 + 12u - 4\ln u = \frac{4}{3}\left(1+\sqrt{z}\right)^3 - 6\left(1+\sqrt{z}\right)^2 + 12\left(1+\sqrt{z}\right) -$$

$$4\ln\left(1+\sqrt{z}\right) = \frac{4}{3}\left(1+\sqrt{\sqrt{x}}\right)^3 - 6\left(1+\sqrt{\sqrt{x}}\right)^2 + 12\left(1+\sqrt{\sqrt{x}}\right) -$$

$4\ln\left(1+\sqrt{\sqrt{x}}\right)$. This expression simplifies to $\frac{4}{3}\sqrt[4]{x^3} - 2\sqrt{x} + 4\sqrt[4]{x} -$

$4\ln\left(1+\sqrt[4]{x}\right) - \frac{14}{3}$.

INTEGRAL 97

Problem:

$$\int \left(\sin^{-1} x \right)^2 dx$$

Solution:

$$x \left(\sin^{-1} x \right)^2 + 2\sqrt{1 - x^2} \, \sin^{-1} x - 2x + constant$$

Techniques used:

Change of variables, Integration by parts, Trigonometric identities

Step-by-step solution:

Let $\sin^{-1} x = z \Rightarrow \sin z = x$ and $dx = \cos z \, dz$. Write the integral in terms of the variable z, as $\int \left(\sin^{-1} x \right)^2 dx = \int z^2 \cos z \, dz$. Using the integration by parts technique, we get $\int \underbrace{z^2}_{f} \underbrace{\cos z}_{dg} dz = z^2 \sin z - 2\int z \sin z \, dz = z^2 \sin z + 2z \cos z - 2 \underbrace{\int \cos z \, dz}_{\sin z}$.

Writing the answer in terms of the original variable x, gives $z^2 \sin z + 2z \cos z - 2 \sin z = x \left(\sin^{-1} x \right)^2 + 2\sqrt{1 - x^2} \, \sin^{-1} x - 2x$.

INTEGRAL 98

Problem:

$$\int e^{x^x} (1 + \ln x) x^{2x} \, dx$$

Solution:

$$e^{x^x} \left(x^x - 1 \right) + constant$$

Techniques used:

Change of variables, Integration by parts

Step-by-step solution:

Let $x^x = z$. Taking logarithm of both sides gives $\ln x^x = \ln z$, or $x \ln x = \ln z$ and after differentiating we have $(1 + \ln x) dx = \dfrac{dz}{z}$.

Therefore, $dx = \dfrac{dz}{x^x (1 + \ln x)}$. Now, write the integral in terms of the variable z, as $\int e^{x^x} (1 + \ln x) x^{2x} \, dx = \int e^z (1 + \ln x) z^2 \dfrac{dz}{z(1 + \ln x)} = $

$\int z e^z dz$. Using the integration by parts technique, the latter integral reads $\int z e^z dz = z e^z - e^z$. This expression in terms of the original variable x gives the answer as $z e^z - e^z = e^{x^x} \left(x^x - 1 \right)$.

INTEGRAL 99

Problem:

$$\int x\pi^{\ln x} dx$$

Solution:

$$\frac{x^2 \pi^{\ln x}}{2 + \ln \pi} + constant$$

Techniques used:

Change of variables, Integration by parts, Logarithmic identities

Step-by-step solution:

Let $\ln x = z \Rightarrow dx = e^z dz$, and write the integral in terms of the variable z, as $\int x\pi^{\ln x} dx = \int e^{2z} \pi^z dz$. Using the integration by

parts, gives $\int \underbrace{e^{2z}}_{dg} \underbrace{\pi^z}_{f} dz = \frac{e^{2z}}{2} \pi^z - \frac{1}{2} \int e^{2z} \left(\frac{d}{dz} \pi^z \right) dz$. For calculating

$\frac{d}{dz} \pi^z$, let $\pi^z = u$ and take the logarithm of both sides. Therefore,

$\ln \pi^z = z \ln \pi = \ln u$. After differentiating both sides, we

get $\frac{d}{dz}(z \ln \pi) = \frac{du}{u}$. Or $du = u \ln \pi = \pi^z \ln \pi$. Note that $du = \frac{d}{dz} \pi^z$.

Back substituting into the latter integral gives $-\frac{1}{2} \int e^{2z} \left(\frac{d}{dz} \pi^z \right) dz =$

$-\frac{1}{2} \int e^{2z} \left(\pi^z \ln \pi \right) dz = -\frac{\ln \pi}{2} \int e^{2z} \pi^z dz$. Now, plugin back into the for-

mer expression, gives $\int e^{2z} \pi^z dz = \frac{e^{2z}}{2} \pi^z - \frac{\ln \pi}{2} \int e^{2z} \pi^z dz$. Rearranging

the terms, gives $\left(1 + \frac{\ln \pi}{2} \right) \int e^{2z} \pi^z dz = \frac{e^{2z}}{2} \pi^z$. Or $\int e^{2z} \pi^z dz = \frac{e^{2z} \pi^z}{2 + \ln \pi}$.

Write this expression in terms of the original variable x, to get the

answer as $\frac{e^{2z} \pi^z}{2 + \ln \pi} = \frac{e^{2 \ln x} \pi^{\ln x}}{2 + \ln \pi} = \frac{x^2 \pi^{\ln x}}{2 + \ln \pi}$. Note that we used the log-

arithmic identity, $e^{2 \ln x} = x^2$.

INTEGRAL 100

Problem:

$$\int \frac{\ln(\sin x)}{1+\sin x}\,dx$$

Solution:

$$2\frac{\ln(\sin x)}{1+\cot(x/2)} - 2\ln\left(\cos\left(\frac{x}{2}\right)\right) = x + \text{constant}$$

Techniques used:

Change of variables, Integration by parts, Trigonometric identities

Step-by-step solution:

Write the sine function as half-angle, using the identity $\sin x = 2\sin\frac{x}{2}\cos\frac{x}{2}$, and expand the integrand to get $\int \frac{\ln(\sin x)}{1+\sin x}\,dx =$

$$\int \frac{\ln\left(2\sin\frac{x}{2}\cos\frac{x}{2}\right)}{1+\sin x}\,dx = (\ln 2)\int \frac{1}{1+\sin x}\,dx + \int \frac{\ln(\sin(x/2))}{1+\sin x}\,dx +$$

$\int \frac{\ln(\cos(x/2))}{1+\sin x}\,dx.$ Now, we have three new integrals to calculate.

For $\int \frac{1}{1+\sin x}\,dx$, write it in terms of half-angle using the identity

$\sin x = \frac{2\tan(x/2)}{1+\tan^2(x/2)}$. Therefore, after some manipulations we get

$$\int \frac{1}{1+\sin x}\,dx = \int \frac{1+\tan^2(x/2)}{(1+\tan(x/2))^2}\,dx. \text{ Let } \tan\left(\frac{x}{2}\right) = z \Rightarrow dx = 2\cos^2$$

$(x/2)dz = \frac{2}{1+z^2}\,dz$, and write the integral in terms of the variable z.

Thus $\int \dfrac{1}{1+\sin x}\,dx = 2\int \dfrac{1}{(1+z)^2}\,dz = \dfrac{-2}{1+z}$. Therefore, the answer in

terms of variable x reads $(\ln 2)\int \dfrac{1}{1+\sin x}\,dx = -\dfrac{2\ln 2}{1+\tan\dfrac{x}{2}} = \dfrac{\ln 1/4}{1+\tan\dfrac{x}{2}}$.

For $\int \dfrac{\ln(\sin x/2)}{1+\sin x}\,dx$, using integration by parts we get

$\int \underbrace{\ln(\sin x/2)}_{f}\,\underbrace{\left(\dfrac{dx}{1+\sin x}\right)}_{dg} = \ln(\sin x/2)\int \dfrac{dx}{1+\sin x} - \int \left(\ln(\sin x/2)\right)'$

$\left(\int \dfrac{dx}{1+\sin x}\right)dx$. But $\left(\ln\left(\sin\dfrac{x}{2}\right)\right)' = \dfrac{d\ln\left(\sin\dfrac{x}{2}\right)}{dx} = \dfrac{\cos x/2}{2\sin x/2}$ and we

have $\int \dfrac{dx}{1+\sin x} = -\dfrac{2}{1+\tan\dfrac{x}{2}}$. Therefore, $\int \dfrac{\ln(\sin x/2)}{1+\sin x}\,dx =$

$\dfrac{-2\ln(\sin x/2)}{1+\tan(x/2)} + \int \dfrac{\cos(x/2)}{\sin(x/2)(1+\tan(x/2))}\,dx$. Write the later

integral in terms of variable z (recall $\tan(x/2) = z$)

as $\int \dfrac{\cos(x/2)}{\sin(x/2)(1+\tan(x/2))}\,dx = 2\int \dfrac{1}{z(1+z)(1+z^2)}\,dz$. For the

latter integral, use partial fraction technique to get

$2\int \dfrac{1}{z(1+z)(1+z^2)}\,dz = 2\int \dfrac{dz}{z} - \int \dfrac{dz}{1+z} - \int \dfrac{dz}{1+z^2} - \int \dfrac{z}{1+z^2}\,dz$. The

first two new integrals from the latter expression read

$2\int \dfrac{dz}{z} = 2\ln z = 2\ln(\tan(x/2))$, and $-\int \dfrac{dz}{1+z} = -\ln(1+z) = -\ln(1+$

$\tan(x/2))$. The remaining integrals can be worked out as

$-\int \dfrac{dz}{1+z^2} = -\tan^{-1} z = -\tan^{-1}(\tan(x/2))$, and $-\int \dfrac{z}{1+z^2}\,dz = -\dfrac{1}{2}$

$\int \dfrac{2z}{1+z^2}\,dz = -\dfrac{1}{2}\ln(1+z^2) = -\dfrac{1}{2}\ln(1+\tan^2(x/2))$.

For $\int \dfrac{\ln(\cos x / 2)}{1+\sin x}dx$, using integration by parts we get

$$\underbrace{\int \ln(\cos x/2)}_{f}\underbrace{\left(\dfrac{dx}{1+\sin x}\right)}_{dg} = \ln(\cos x/2)\int \dfrac{dx}{1+\sin x} - \int \left(\ln(\cos x/2)\right)'$$

$\left(\int \dfrac{dx}{1+\sin x}\right)dx.$ But $\left(\ln(\cos x/2)\right)' = \dfrac{d\ln(\cos x/2)}{dx} = \dfrac{-\sin x/2}{2(\cos x/2)}$

and we have $\int \dfrac{dx}{1+\sin x} = -\dfrac{2}{1+\tan\dfrac{x}{2}}$. Therefore, $\int \dfrac{\ln(\cos x/2)}{1+\sin x}dx =$

$\dfrac{-2\ln(\cos x/2)}{1+\tan(x/2)} - \int \dfrac{\sin(x/2)}{(\cos x/2)(1+\tan(x/2))}dx.$ Write the latter

integral in terms of variable z (recall $\tan(x/2)=z$) as

$-\int \dfrac{\sin(x/2)}{(\cos(x/2))(1+\tan(x/2))}dx = -2\int \dfrac{z}{(1+z)(1+z^2)}dz.$ For the

latter integral, we use partial fraction technique to get

$-2\int \dfrac{z}{(1+z)(1+z^2)}dz = \int \dfrac{dz}{1+z} - \int \dfrac{z}{1+z^2}dz - \int \dfrac{1}{1+z^2}dz.$ The first two

new integrals from the latter expression read $\int \dfrac{dz}{1+z} =$

$\ln(1+z) = \underline{\ln\left(1+\tan(x/2)\right)},$ and $-\int \dfrac{z}{1+z^2}dz = -(1/2)\ln\left(1+z^2\right) =$

$\underline{-(1/2)\ln\left(1+\tan^2(x/2)\right)}.$ The remaining integral can be worked

out as $-\int \dfrac{dz}{1+z^2} = -\tan^{-1}z = \underline{-\tan^{-1}\left(\tan(x/2)\right)=-x/2}.$

Collection all related underlined answers, we get the solution to the integral, after some simplifications, as $2\dfrac{\ln(\sin x)}{1+\cot(x/2)} -$

$2\ln\left(\cos\left(\dfrac{x}{2}\right)\right) - x.$

PART 2

EXAMPLES APPLIED IN ENGINEERING

In this part of the book, we present some integrals along with their solutions related to engineering topics. The list is not exclusive but meant to help readers with their learning from Part 1 with some application examples in technical computations. The examples will focus on area properties of sections, structural beams, and a couple of probability distributions commonly used in engineering fields.

1 SEMI-CIRCLE SHAPES

We consider a half-circle shape with radius R as shown in Figure 1. Its area, centroid, and moment of inertia is calculated using related integrals and the worked-out solutions.

A half-circle shape can be the cross section of a beam, rod, or shape of hydraulic gate, for example.

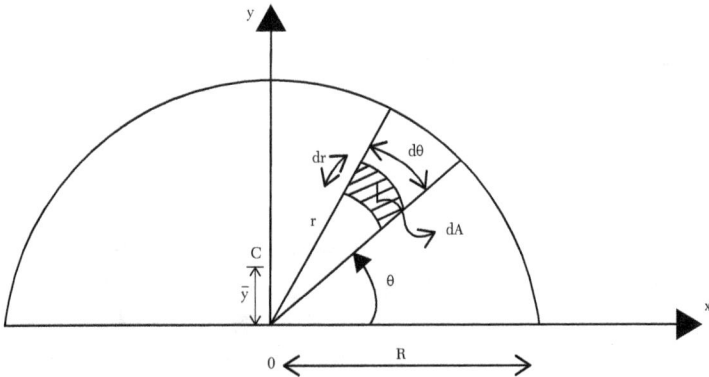

Figure 1 A Semi-circle shape and differential area element

<u>Area</u> is equal to $\dfrac{\pi R^2}{2}$, from geometry. We calculate the area using integration of area differential element $dA = r\,dr\,d\theta$, where r is the radial disctance from the center and θ is polar angle w.r.t x-axis. Therefore, $A = \int\limits_0^\pi\int\limits_0^R r\,dr\,d\theta = \int\limits_0^\pi d\theta \int\limits_0^R r\,dr = \int\limits_0^\pi \left(\dfrac{R^2}{2}\right)d\theta = \dfrac{R^2}{2}\int\limits_0^\pi d\theta = \underline{\dfrac{\pi R^2}{2}}$.

<u>Centroid</u> is the first moment of area divided by its total area and is located at $\left(0, \bar{y}\right)$ when symmetry about y-axis exists as shown in Figure 1. Or $\bar{y} = \dfrac{\int y\,dA}{\int dA} = \dfrac{\int y\,dA}{A}$, where $y = r\sin\theta$. Note that y is measured from the centroid of the differential element. But $\int y\,dA = \int\limits_0^\pi\int\limits_0^R r^2\sin\theta\,dr\,d\theta = \int\limits_0^\pi \sin\theta\,d\theta \int\limits_0^R r^2\,dr = \int\limits_0^\pi \left(\dfrac{R^3}{3}\right)\sin\theta\,d\theta = -\dfrac{R^3}{3}\cos\theta\Big|_0^\pi = -\dfrac{R^3}{3}(-1-1) = \dfrac{2R^3}{3}$. Therefore, $\bar{y} = \dfrac{\int y\,dA}{\pi R^2/2} = \dfrac{2R^3}{3}\dfrac{2}{\pi R^2} = \underline{\dfrac{4R}{3\pi}}$.

<u>Moment of inertia</u> is the second moment of area with reference to a desired axis. Considering x-axis as reference,

$$I_x = \int y^2 dA = \int_0^\pi \int_0^R r^3 \sin^2 \theta \, dr d\theta = \int_0^\pi \sin^2 \theta \, d\theta \int_0^R r^3 dr = \frac{R^4}{4} \int_0^\pi \left(\frac{1-\cos 2\theta}{2} \right) d\theta.$$

Note that y is measured from the centroid of the differential element. Performing the integrations gives $\dfrac{R^4}{4} \displaystyle\int_0^\pi \left(\dfrac{1-\cos 2\theta}{2} \right) d\theta =$

$$\frac{R^4}{8} \left[\left(\theta - \frac{\sin 2\theta}{2} \right) \right]_0^\pi = \frac{\pi R^4}{8}. \text{ Note that } I_y = I_x = \frac{\pi R^4}{8}. \text{ We can calcu-}$$

late $I_y = \int x^2 dA = \displaystyle\int_0^\pi \int_0^R r^3 \cos^2 \theta \, dr d\theta$, directly as well.

Using the parallel axis theorem, we can calculate the moment of inertia with respect to the axis at the centroid, I_c. Or $I_x = I_c + Ad^2$, where d is the normal distance between x-axis and that parallel passing through the point C at the centroid, or $d = \bar{y}$. Therefore

$$I_c = I_x - Ad^2 = \frac{\pi R^4}{8} - \frac{\pi R^2}{2} \left(\frac{4R}{3\pi} \right)^2 = \underline{\left(\frac{9\pi^2 - 64}{72\pi} \right) R^4}.$$

<u>Polar moment of inertia</u> is the second moment of area with reference to a desired point. Considering the origin as the reference,

$$J_o = \int r^2 dA = \int_0^\pi \int_0^R r^3 dr d\theta = \int_0^\pi d\theta \int_0^R r^3 dr = \frac{R^4}{4} \int_0^\pi d\theta = \underline{\frac{\pi R^4}{4}}. \quad \text{Note that}$$

$$J_o = I_x + I_y.$$

Using the parallel axis theorem, we can calculate the polar moment of inertia with respect to the centroid, J_c. Or $J_o = J_c + Ad^2$, where d is the distance between origin and the centroid. Therefore,

$$J_c = J_o - Ad^2 = \frac{\pi R^4}{4} - \frac{\pi R^2}{2} \left(\frac{4R}{3\pi} \right)^2 = \underline{\left(\frac{9\pi^2 - 32}{36\pi} \right) R^4}.$$

Table 2 lists the results for semi-circle shape.

TABLE 2 Results for Semi-circle

Shape	Area	Centroid	I_x	I_c	J_o	J_c
Semi-circle	$\dfrac{\pi R^2}{2}$	$\dfrac{4R}{3\pi}$	$\dfrac{\pi R^4}{8}$	$\left(\dfrac{9\pi^2 - 64}{72\pi} \right) R^4$	$\dfrac{\pi R^4}{4}$	$\left(\dfrac{9\pi^2 - 32}{36\pi} \right) R^4$

2   CIRCULAR SEGMENT SHAPES

We consider a segmental shape of a half circle with radius R as shown in Figure 2, shaded area. Its area, centroid, and moment of inertia is calculated using related integrals and the worked-out solutions.

A circular segmental shape can be the cross section of a beam, rod, or shape of hydraulic gate, for example.

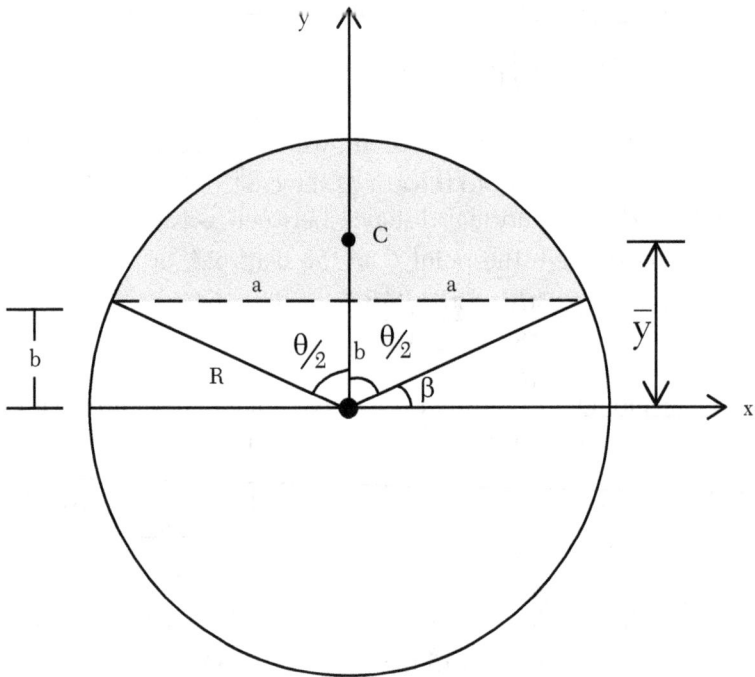

Figure 2 A Circular segment area shape

Assuming total angle θ, from geometry we have $a = R \sin \dfrac{\theta}{2}$, $b = R \cos \dfrac{\theta}{2}$, and $\beta = \dfrac{\pi - \theta}{2}$. The equation of a circle reads $x^2 + y^2 = R^2$. Therefore, for the sector we can write $y = R \sqrt{1 - \dfrac{x^2}{R^2}}$.

<u>Area:</u> We calculate the area using the integration of area differential

element $dA = dxdy$. Therefore, $A = \int\int dxdy = \int_{b}^{y} dy \int_{-a}^{a} dx = \int_{-a}^{a}(y-b)$

$dx = 2\int_{0}^{a}(y-b)dx = 2\int_{0}^{a} ydx - 2b\int_{0}^{a} dx = 2\int_{0}^{a} ydx - 2ab.$ But $2\int_{0}^{a} ydx = 2R\int_{0}^{a}$

$\sqrt{1-\dfrac{x^2}{R^2}}\,dx.$ Let $\dfrac{x}{R} = \cos\beta \Rightarrow dx = -R\sin\beta\,d\beta.$ Therefore, the

limits of the integral reads $x = \left(0, a = R\sin\dfrac{\theta}{2}\right) \equiv \beta = \left(\dfrac{\pi}{2}, \dfrac{\pi-\theta}{2}\right)$

and write the latter integral in terms of the variable β to get

$2R\int_{0}^{a}\sqrt{1-\dfrac{x^2}{R^2}}\,dx = -2R^2\int_{\frac{\pi}{2}}^{\frac{\pi-\theta}{2}} \sin^2\beta\,d\beta = -R^2\int_{\frac{\pi}{2}}^{\frac{\pi-\theta}{2}}(1-\cos 2\beta)\,d\beta =$

$\left[\dfrac{\sin 2\beta}{2} - \beta\right]_{\frac{\pi}{2}}^{\frac{\pi-\theta}{2}}.$ After applying the limits, we get $R^2\left[\dfrac{\sin 2\beta}{2} - \beta\right]_{\frac{\pi}{2}}^{\frac{\pi-\theta}{2}} =$

$R^2\left(\dfrac{\overbrace{\sin(\pi-\theta)}^{=\sin\theta}}{2} - \dfrac{\pi-\theta}{2} - \dfrac{\sin\pi}{2} + \dfrac{\pi}{2}\right) = \dfrac{R^2}{2}(\theta + \sin\theta).$ Collecting all

related answers, we get $A = \dfrac{R^2}{2}(\theta + \sin\theta) - 2ab = \dfrac{R^2}{2}(\theta + \sin\theta) -$

$\underbrace{2R^2\sin\dfrac{\theta}{2}\cos\dfrac{\theta}{2}}_{R^2\sin\theta}.$ Hence, $\underline{A = \dfrac{R^2}{2}(\theta - \sin\theta).}$

<u>Centroid</u> is the first moment of area divided by its total area

and is located at $C(0,\bar{y})$ when symmetry about y-axis exists as

shown in Figure 2. Or $\bar{y} = \dfrac{\int ydA}{\int dA} = \dfrac{\int ydA}{A}$, where y is measured

from the x-axis to the centroid of the differential element. But

$\int ydA = \int_{b}^{y} ydy \int_{-a}^{a} dx = \dfrac{1}{2}\int_{-a}^{a}(y^2 - b^2)dx = \int_{0}^{a} y^2 dx - ab^2.$

The latter integral reads $\int_0^a y^2 dx = \int_0^a (R^2 - x^2) dx = \left[R^2 x - \dfrac{x^3}{3} \right]_0^a =$

$aR^2 - \dfrac{a^3}{3}$. Therefore, $\int_0^a y^2 dx - ab^2 = aR^2 - \dfrac{a^3}{3} - ab^2 = R^3 \sin\dfrac{\theta}{2} -$

$\dfrac{R^3 \sin^3 \dfrac{\theta}{2}}{3} - R^3 \sin\dfrac{\theta}{2} \underbrace{\cos^2 \dfrac{\theta}{2}}_{1 - \sin^2 \frac{\theta}{2}} = \dfrac{2}{3} R^3 \sin^3 \dfrac{\theta}{2}$. Now, we have $\bar{y} =$

$\dfrac{\dfrac{2}{3} R^3 \sin^3 \dfrac{\theta}{2}}{\dfrac{R^2}{2}(\theta - \sin\theta)} = \dfrac{4R \sin^3 \left(\dfrac{\theta}{2} \right)}{3(\theta - \sin\theta)}.$

<u>Moment of inertia</u> is the second moment of area with reference to a desired axis. Considering x-axis as reference,

$I_x = \int y^2 dA = \int_b^y y^2 dy \int_{-a}^a dx = \dfrac{2}{3} \int_0^a (y^3 - b^3) dx = \dfrac{2}{3} \int_0^a y^3 dx - \dfrac{2}{3} ab^3$. But

$-\dfrac{2}{3} ab^3 = -\dfrac{2}{3} R^4 \sin^3 \dfrac{\theta}{2} \cos^3 \dfrac{\theta}{2} = -\dfrac{R^4}{3} \sin\theta \left(1 - \sin^2 \dfrac{\theta}{2} \right)$. Now, after sub-

stituting for y, the latter integral reads $\dfrac{2}{3} \int_0^a y^3 dx = \dfrac{2R^3}{3} \int_0^a \left(\sqrt{1 - \dfrac{x^2}{R^2}} \right)^3 dx$.

After rewriting the latter integral in terms of the variable β (recall $x = R\cos\beta$), we get $\dfrac{2R^3}{3} \int_0^a \left(\sqrt{1 - \dfrac{x^2}{R^2}} \right)^3 dx = -\dfrac{2R^4}{3}$

$\int_{\frac{\pi}{2}}^{\frac{\pi - \theta}{2}} \sqrt{(1 - \cos^2 \beta)^3} \sin\beta \, d\beta = -\dfrac{2R^4}{3} \int_{\frac{\pi}{2}}^{\frac{\pi - \theta}{2}} \sin^4 \beta \, d\beta$. But using the

trigonometric relation, $\sin^4 \beta = \dfrac{1}{8}(3 - 4\cos 2\beta + \cos 4\beta)$ we get

$-\dfrac{R^4}{12} \int_{\frac{\pi}{2}}^{\frac{\pi - \theta}{2}} (3 - 4\cos 2\beta + \cos 4\beta) d\beta = -\dfrac{R^4}{12} \left[3\beta - 2\sin 2\beta + \dfrac{1}{4}\sin 4\beta \right]_{\frac{\pi}{2}}^{\frac{\pi - \theta}{2}} =$

$\dfrac{R^4}{12}\left(\dfrac{3}{2}\theta + 2\sin\theta + \dfrac{1}{4}\sin 2\theta\right)$. But we have $\dfrac{1}{4}\sin 2\theta = \dfrac{1}{2}\sin\theta\cos\theta =$

$\dfrac{1}{2}\sin\theta\left(1 - 2\sin^2\dfrac{\theta}{2}\right) = \dfrac{1}{2}\sin\theta - \sin\theta\sin^2\dfrac{\theta}{2}$. Therefore, $\dfrac{R^4}{12}\left(\dfrac{3}{2}\theta +\right.$

$\left.2\sin\theta + \dfrac{1}{4}\sin 2\theta\right) = \dfrac{R^4}{12}\left(\dfrac{3}{2}\theta + \dfrac{5}{2}\sin\theta - \sin\theta\sin^2\dfrac{\theta}{2}\right)$. Collecting the

results, we get $I_x = \dfrac{R^4}{12}\left(\dfrac{3}{2}\theta + \dfrac{5}{2}\sin\theta - \sin\theta\sin^2\dfrac{\theta}{2}\right) - \dfrac{R^4}{3}\sin\theta\left(1 -\right.$

$\left.\sin^2\dfrac{\theta}{2}\right)$. After simplification, we get $I_x = \dfrac{R^4}{8}\left(\theta - \sin\theta + 2\sin\theta\sin^2\dfrac{\theta}{2}\right)$.

Similarly, considering y-axis as reference, $I_y = \displaystyle\int x^2 dA = \int_b^y dy\int_{-a}^a x^2 dx =$

$2\displaystyle\int_0^a (y - b)x^2 dx = 2\int_0^a yx^2 dx - 2b\int_0^a x^2 dx$. But $-2b\displaystyle\int_0^a x^2 dx = -\dfrac{2a^3 b}{3} = -$

$\dfrac{2}{3}R^4\sin^3\dfrac{\theta}{2}\cos\dfrac{\theta}{2} = -\dfrac{R^4}{3}\sin\theta\sin^2\dfrac{\theta}{2}$. Now, after substituting for y,

the latter integral reads $2\displaystyle\int_0^a yx^2 dx = 2R\int_0^a x^2\sqrt{1 - \dfrac{x^2}{R^2}}\, dx$. After rewrit-

ing the latter integral in terms of variable β (recall $x = R\cos\beta$),

we get $2R\displaystyle\int_0^a x^2\sqrt{1 - \dfrac{x^2}{R^2}}\, dx = -2R^4\int_{\frac{\pi}{2}}^{\frac{\pi - \theta}{2}} \sin^2\beta\cos^2\beta\, d\beta$. But using

the trigonometric relation, $\sin^2\beta\cos^2\beta = \dfrac{1}{8}(1 - \cos 4\beta)$ we get

$-2R^4\displaystyle\int_{\frac{\pi}{2}}^{\frac{\pi - \theta}{2}} \sin^2\beta\cos^2\beta\, d\beta = -\dfrac{R^4}{4}\int_{\frac{\pi}{2}}^{\frac{\pi - \theta}{2}} (1 - \cos 4\beta)\, d\beta = \dfrac{R^4}{4}\left[-\beta +\right.$

$\dfrac{1}{4}\sin 4\beta\Big]_{\frac{\pi}{2}}^{\frac{\pi - \theta}{2}} = \dfrac{R^4}{4}\left(\dfrac{\theta}{2} - \dfrac{1}{4}\sin 2\theta\right)$. But we have $\dfrac{1}{4}\sin 2\theta =$

$\dfrac{1}{2}\sin\theta\cos\theta = \dfrac{1}{2}\sin\theta\left(1 - 2\sin^2\dfrac{\theta}{2}\right) = \dfrac{1}{2}\sin\theta - \sin\theta\sin^2\dfrac{\theta}{2}$. Therefore,

$$\frac{R^4}{4}\left(\frac{\theta}{2}-\frac{1}{4}\sin 2\theta\right)=\frac{R^4}{4}\left(\frac{\theta}{2}-\frac{1}{2}\sin\theta+\sin\theta\sin^2\frac{\theta}{2}\right).$$ Collecting the

results, after some simplifications, we get $I_y=\dfrac{R^4}{24}\Big(3\theta-3\sin\theta-$

$2\sin\theta\sin^2\dfrac{\theta}{2}\Big).$

Using the parallel axis theorem, we can calculate the moment of inertia with respect to the centroid, I_c. Or $I_x=I_c+Ad^2$, where d is the distance between x-axis and the centroid, or $\bar{y}$. Therefore

$$I_c=I_x-Ad^2=\frac{R^4}{8}\left(\theta-\sin\theta+2\sin\theta\sin^2\frac{\theta}{2}\right)-\frac{R^2}{2}(\theta-\sin\theta)$$

$$\left[\frac{4R\sin^3\left(\dfrac{\theta}{2}\right)}{3(\theta-\sin\theta)}\right]^2.$$ Or, after simplification, $I_c=\dfrac{R^4}{8}\Bigg[\theta-\sin\theta+$

$$2\sin\theta\sin^2\left(\frac{\theta}{2}\right)-\frac{64}{9}\cdot\frac{\sin^6\left(\dfrac{\theta}{2}\right)}{\theta-\sin\theta}\Bigg]$$

<u>Polar moment of inertia</u> is the second moment of area with reference to a desired point. Considering the origin as the reference, $J_o=\int r^2 dA=\int(x^2+y^2)dA=I_x+I_y.$ Therefore, using previously obtained results, we have

$$J_o=\frac{R^4}{8}\left(\theta-\sin\theta+2\sin\theta\sin^2\frac{\theta}{2}\right)+\frac{R^4}{24}\left(3\theta-3\sin\theta-2\sin\theta\sin^2\frac{\theta}{2}\right).$$

After some simplifications, we have $J_o=\dfrac{R^4}{4}\Big(\theta-\sin\theta+\dfrac{2}{3}\sin\theta$

$\sin^2\dfrac{\theta}{2}\Big).$

Using the parallel axis theorem, we can calculate the polar moment of inertia with respect to the centroid, J_c. Or $J_o = J_c + Ad^2$, where d is the distance between origin and the centroid, or $\bar{y}$. Therefore

$$J_c = J_o - Ad^2 = \frac{R^4}{4}\left(\theta - \sin\theta + \frac{2}{3}\sin\theta\sin^2\frac{\theta}{2}\right) - \frac{R^2}{2}(\theta - \sin\theta)$$

$$\left[\frac{4R\sin^3\left(\dfrac{\theta}{2}\right)}{3(\theta - \sin\theta)}\right]^2 .$$

Table 3 lists some of the results for circular segment shape.

TABLE 3 Results for circular segment

Shape	Area	Centroid	I_x	J_o
Circular segment	$\dfrac{R^2}{2}(\theta - \sin\theta)$	$\dfrac{4R\sin^3\left(\dfrac{\theta}{2}\right)}{3(\theta - \sin\theta)}$	$\dfrac{R^4}{8}\left(\theta - \sin\theta + 2\sin\theta\sin^2\dfrac{\theta}{2}\right)$	$\dfrac{R^4}{4}\left(\theta - \sin\theta + \dfrac{2}{3}\sin\theta\sin^2\dfrac{\theta}{2}\right)$

3   SEMI-ELLIPSE SHAPES

We consider a half-ellipse shape with major and minor radii a and b, respectively. Its area, centroid, and moment of inertia is calculated using integrals.

A half-ellipse shape can be the cross-section shape of beams, rods, or hydraulic gates for example.

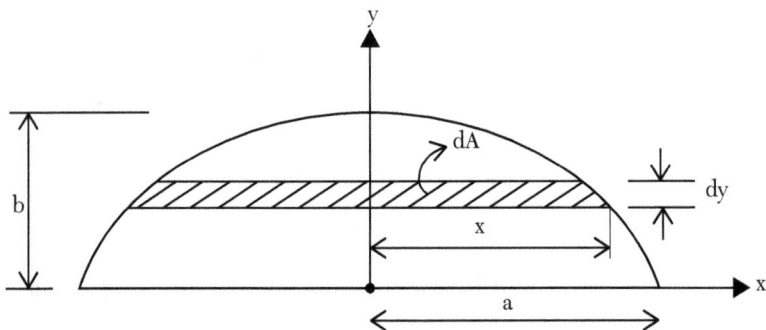

Figure 3 A Semi ellipse area shape and differential element

Area is equal to $\dfrac{\pi ab}{2}$, from geometry. We calculate the area using integration of the area differential element $dA = 2x\,dy$ as shown in Figure 3. Using the equation of ellipse, $\dfrac{x^2}{a^2} + \dfrac{y^2}{b^2} = 1$ we have

$$A = 2\int x\,dy = 2a\int_0^b \sqrt{1 - \frac{y^2}{b^2}}\,dy. \quad \text{Let} \quad \frac{y}{b} = \sin\theta \Rightarrow dy = b\cos\theta\,d\theta \quad \text{and}$$

$$A = 2ab\int_0^{\frac{\pi}{2}} \sqrt{1 - \sin^2\theta}\,\cos\theta\,d\theta = 2ab\int_0^{\frac{\pi}{2}} \cos^2\theta\,d\theta = ab\int_0^{\frac{\pi}{2}} (1 + \cos 2\theta)\,d\theta.$$

Performing the integration gives $ab\left[\theta + \dfrac{\sin 2\theta}{2}\right]_0^{\pi/2} = \dfrac{\pi ab}{2}$. Note that due to symmetry, area of an ellipse is πab.

Centroid is the first moment of area divided by its total area and is located at $(0, \bar{y})$ when symmetry about y-axis exists.

Or $\quad \bar{y} = \dfrac{\int y dA}{\int dA} = \dfrac{\int y dA}{A}.\quad$ But $\quad \int y dA = 2\int\limits_0^b xydy = 2a\int\limits_0^b y\sqrt{1-\dfrac{y^2}{b^2}}dy.$

Performing the integration, after writing it in terms of the variable θ, for values $y=(0,b)\equiv\theta=\left(0,\dfrac{\pi}{2}\right)$ gives,

$$2a\int\limits_0^b y\sqrt{1-\dfrac{y^2}{b^2}}dy = 2ab^2\int\limits_0^{\frac{\pi}{2}}\sin\theta\cos^2\theta d\theta = 2ab^2\left[-\dfrac{\cos^3\theta}{3}\right]_0^{\frac{\pi}{2}} = \dfrac{2ab^2}{3}.$$

Therefore, $\bar{y} = \dfrac{2ab^2/3}{\pi ab/2} = \dfrac{4b}{3\pi}.$

<u>Moment of inertia</u> is the second moment of area with reference to a desired axis. Considering the x-axis as reference,

$$I_x = \int y^2 dA = 2\int\limits_0^b y^2 xdy = 2a\int\limits_0^b y^2\sqrt{1-\dfrac{y^2}{b^2}}dy = 2ab^3\int\limits_0^{\frac{\pi}{2}}\sin^2\theta\cos^2\theta d\theta.$$

Performing the integrations gives $\quad\dfrac{ab^3}{2}\int\limits_0^{\frac{\pi}{2}}\sin^2 2\theta d\theta = \dfrac{ab^3}{4}$

$$\int\limits_0^{\frac{\pi}{2}}(1-\cos 4\theta)d\theta = \dfrac{ab^3}{4}\left[\left(\theta-\dfrac{\sin 4\theta}{4}\right)\right]_0^{\frac{\pi}{2}} = \dfrac{\pi ab^3}{8} = I_x.\quad$$ Similarly, we

can calculate $I_y = \int x^2 dA = \dfrac{\pi a^3 b}{8}.$ Note that x is measured from the centroid of the differential element.

Using the parallel axis theorem, we can calculate the moment of inertia with respect to the axis at the centroid, I_{cx}. Or $I_x = I_{cx} + Ad^2$, where d is the normal distance between axis x and that parallel passing through the centroid, or $\bar{y}$. Therefore

$$I_{cx} = I_x - Ad^2 = \dfrac{\pi ab^3}{8} - \dfrac{\pi ab}{2}\left(\dfrac{4b}{3\pi}\right)^2 = \left(\dfrac{9\pi^2-64}{72\pi}\right)ab^3.\text{ Similarly,}$$

$$I_{cy} = I_y - Ad^2 = \dfrac{\pi a^3 b}{8} - \dfrac{\pi ab}{2}\left(\dfrac{4b}{3\pi}\right)^2 = \left(\dfrac{9\pi^2-64}{72\pi}\right)a^3 b.$$

Polar moment of inertia is the second moment of area with reference to a desired point. Considering the origin as the reference,

$$J_o = \int r^2 dA = \int (x^2 + y^2) dA = I_x + I_y = \frac{\pi ab^3}{8} + \frac{\pi a^3 b}{8} = \frac{\pi ab}{8}(a^2 + b^2).$$

we can calculate the moment of inertia with respect to the centroid, J_c. Or $J_c = J_{cx} + J_{cy}$. Therefore, $J_o = \left(\frac{9\pi^2 - 64}{72\pi} \right)(a^3 b + ab^3)$.

Table 4 lists the results for semi-ellipse shape.

Table 4 Results for semi-ellipse

Shape	Area	Centroid	Ix	Icx	Jo	Jc
Semi-ellipse	$\dfrac{\pi ab}{2}$	$\dfrac{4b}{3\pi}$	$\dfrac{\pi ab^3}{8}$	$\left(\dfrac{9\pi^2 - 64}{72\pi} \right) ab^3$	$\dfrac{\pi ab}{8}(a^2 + b^2)$	$\left(\dfrac{9\pi^2 - 64}{72\pi} \right)(a^3 b + ab^3)$

4   TWO-DEGREE POLYNOMIAL SHAPE-QUADRATIC

Consider a quadratic polynomial shape of degree two, $y = b\left(1 - \dfrac{x^2}{a^2}\right)$

which is bounded between coordinate axes in the first quadrant (i.e., $x > 0$ and $y > 0$) as shown in Figure 4.

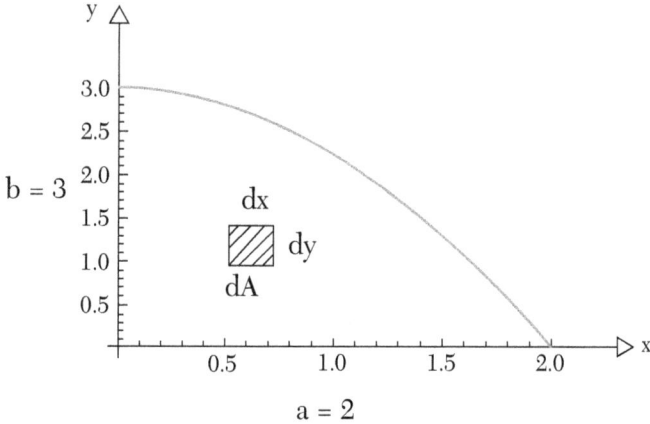

Figure 4 A quadratic area shape with a=2 and b=3.

<u>Area,</u> we calculate the area using integration of the differential area element $dA = dxdy$. Using the equation of polynomial, we have $A = \int dA = \int_0^y dy \int_0^a dx = \int_0^a y\,dx = b \int_0^a \left(1 - \dfrac{x^2}{a^2}\right) dx.$ Performing the integration gives $A = b\left[x - \dfrac{x^3}{3a^2}\right]_0^a = \dfrac{2ab}{3}.$ Simply, subtracting this area from that of the enclosing rectangle, ab we get the value of the area over the polynomial, or $\dfrac{ab}{3}.$

<u>Centroid</u> is the first moment of area divided by its total area. Or $\bar{y}_c = \dfrac{\int y\,dA}{\int dA} = \dfrac{\int y\,dA}{A}.$ But $\int y\,dA = \int_0^y y\,dy \int_0^a dx = \dfrac{1}{2}\int_0^a y^2\,dx = \dfrac{b^2}{2}$

$$\int_0^a \left(1 - \frac{x^2}{a^2}\right)^2 dx = \frac{b^2}{2}\int_0^a\left(1 + \frac{x^4}{a^4} - 2\frac{x^2}{a^2}\right)dx.$$ Performing the integration,

gives, $\dfrac{b^2}{2}\left[x + \dfrac{x^5}{5a^4} - \dfrac{2x^3}{3a^2}\right]_0^a = \dfrac{4ab^2}{15}.$ Therefore, $\bar{y}_c = \dfrac{4ab^2/15}{2ab/3} = \underline{\dfrac{2b}{5}}.$

Similarly, we can calculate the x-coordinate of the centroid,

$$\bar{x}_c = \frac{\int xdA}{\int dA} = \frac{\int xdA}{A}.$$ But $\int xdA = \int_0^y dy \int_0^a xdx = \int_0^a yxdx = b\int_0^a x\left(1 - \frac{x^2}{a^2}\right)dx =$

$b\int_0^a \left(x - \dfrac{x^3}{a^2}\right)dx.$ Performing the integration, gives, $b\left[\dfrac{x^2}{2} - \dfrac{x^4}{4a^2}\right]_0^a =$

$\dfrac{a^2b}{4}.$ Therefore, $\bar{x}_c = \dfrac{a^2b/4}{2ab/3} = \underline{\dfrac{3a}{8}}.$

<u>Moment of inertia</u> is the second moment of area with reference to a desired axis. Considering x-axis as reference,

$$I_x = \int y^2 dA = \int\int y^2 dxdy = \int_0^y y^2 dy\int_0^a dx = \frac{1}{3}\int_0^a y^3 dx = \frac{b^3}{3}\int_0^a\left(1 - \frac{x^2}{a^2}\right)^3 dx = \frac{b^3}{3}$$

$$\int_0^a\left(1 - \frac{x^6}{a^6} + \frac{3x^4}{a^4} - \frac{3x^2}{a^2}\right)dx.$$ Performing the integrations gives

$I_x = \dfrac{b^3}{3}\left[x - \dfrac{x^7}{7a^6} + \dfrac{3x^5}{5a^4} - \dfrac{x^3}{a^2}\right]_0^a = \underline{\dfrac{16ab^3}{105}}.$ Similarly, we can calcu-

late $I_y = \int x^2 dA = \int\int x^2 dxdy = \int_0^y dy\int_0^a x^2 dx = \int_0^a yx^2 dx = b\int_0^a\left(x^2 - \dfrac{x^4}{a^2}\right)dx.$

Performing the integrations gives $I_y = b\left[\dfrac{x^3}{3} - \dfrac{x^5}{5a^2}\right]_0^a = \underline{\dfrac{2a^3b}{15}}.$

Using the parallel axis theorem, we can calculate the moment of inertia with respect to the axis at the centroid, $I_{cx}.$ Or $I_x = I_{cx} + Ad^2$, where d is the normal distance between axis x and that parallel passing through the centroid, or $\bar{y}_c.$

Therefore $I_{cx} = I_x - A\bar{y}_c^2 = \dfrac{16ab^3}{105} - \dfrac{2ab}{3}\left(\dfrac{2b}{5}\right)^2 = \dfrac{8ab^3}{175}$. Similarly,

$I_{cy} = I_y - A\bar{x}_c^2 = \dfrac{2a^3b}{15} - \dfrac{2ab}{3}\left(\dfrac{3a}{8}\right)^2 = \dfrac{19}{480}a^3b.$

<u>Polar moment of inertia</u> is the second moment of area with reference to a desired point. Considering the origin,

$J_o = \int r^2 dA = \int \left(x^2 + y^2\right) dA = I_x + I_y = \dfrac{16ab^3}{105} + \dfrac{2a^3b}{15} = \dfrac{2ab}{105}\left(7a^2 + 8b^2\right).$

we can calculate the polar moment of inertia with respect to the

centroid, J_c. Or $J_c = \dfrac{8ab^3}{175} + \dfrac{19a^3b}{480} = \dfrac{ab}{16800}\left(665a^2 + 768b^2\right).$

Table 5 lists the results for a quadratic/parabolic polynomial shape.

TABLE 5 Results for a parabolic shape

Shape	Area	Centroid, yc	Ix	Icx	Jo	Jc
Semi-ellipse	$\dfrac{2ab}{3}$	$\dfrac{2b}{5}$	$\dfrac{16ab^3}{105}$	$\dfrac{8ab^3}{175}$	$\dfrac{2ab}{105}\left(7a^2 + 8b^2\right)$	$a\dfrac{ab}{16800}\left(665a^2 + 768b^2\right)$

5   THREE-DEGREE POLYNOMIAL SHAPE-CUBIC

Consider a cubic polynomial shape of degree three, $y = b\left(1 - \dfrac{x^3}{a^3}\right)$ which is bounded between coordinate axes in the first quadrant (i.e., $x > 0$ and $y > 0$) as shown in Figure 5

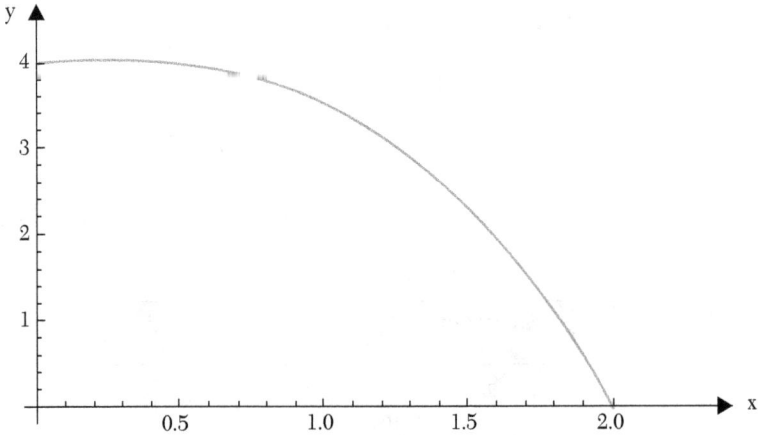

Figure 5 A cubic area shape with a=2 and b=4

Area, we calculate the area using integration of the differential area element $dA = dxdy$. Using the equation of polynomial, we have $A = \int dxdy = \int\limits_0^y dy \int\limits_0^a dx = \int\limits_0^a ydx = b\int\limits_0^a\left(1 - \dfrac{x^3}{a^3}\right)dx$. Performing the integration gives $A = b\left[x - \dfrac{x^4}{4a^3}\right]_0^a = \dfrac{3ab}{4}$. Simply, subtracting this area from that of the enclosing rectangle, ab we get the value of the area over the polynomial, or $\dfrac{ab}{4}$.

Centroid is the first moment of area divided by its total area. Or $\bar{y}_c = \dfrac{\int ydA}{\int dA} = \dfrac{\int ydA}{A}$. But $\int ydA = \int\limits_0^y ydy \int\limits_0^a dx = \dfrac{1}{2}\int\limits_0^a y^2 dx = \dfrac{b^2}{2}$

$$\left(1-\frac{x^3}{a^3}\right)^2 dx = \frac{b^2}{2}\int_0^a\left(1+\frac{x^6}{a^6}-2\frac{x^3}{a^3}\right)dx.$$ Performing the integration,

gives, $\dfrac{b^2}{2}\left[x+\dfrac{x^7}{7a^6}-\dfrac{2x^4}{4a^3}\right]_0^a = \dfrac{9ab^2}{28}.$ Therefore, $\bar{y}_c = \dfrac{9ab^2/28}{3ab/4} = \dfrac{3b}{7}.$

Similarly, we can calculate the x-coordinate of the centroid, $\bar{x}_c = \dfrac{\int x dA}{\int dA} = \dfrac{\int x dA}{A}.$ But $\int x dA = \int_0^y dy\int_0^a x dx = \int_0^a yx dx =$

$b\int_0^a x\left(1-\dfrac{x^3}{a^3}\right)dx = b\int_0^a\left(x-\dfrac{x^4}{a^3}\right)dx.$ Performing the integration, gives,

$b\left[\dfrac{x^2}{2}-\dfrac{x^5}{5a^3}\right]_0^a = \dfrac{3a^2b}{10}.$ Therefore, $\bar{x}_c = \dfrac{3a^2b/10}{3ab/4} = \dfrac{2a}{5}.$

Moment of inertia is the second moment of area with reference to a desired axis. Considering x-axis as reference,

$$I_x = \int y^2 dA = \int\int y^2 dxdy = \int_0^a y^2 dy = \int dx = \frac{1}{3}\int_0^a y^3 dx = \frac{b^3}{3}\int_0^a\left(1-\frac{x^3}{a^3}\right)^3 dx =$$

$$\frac{b^3}{3}\int_0^a\left(1-\frac{x^9}{a^9}+\frac{3x^6}{a^6}-\frac{3x^3}{a^3}\right)^3 dx.$$ Performing the integrations gives

$I_x = \dfrac{b^3}{3}\left[x-\dfrac{x^{10}}{10a^9}+\dfrac{3x^7}{7a^6}-\dfrac{3x^4}{4a^3}\right]_0^a = \dfrac{81ab^3}{140}.$ Similarly, we can calcu-

late $I_y = \int x^2 dA = \int\int x^2 dxdy = \int_0^y dy\int_0^a x^2 dx = \int_0^a yx^2 dx = b\int_0^a\left(x^2-\dfrac{x^5}{a^3}\right)dx.$

Performing the integrations gives $I_y = b\left[\dfrac{x^3}{3}-\dfrac{x^6}{6a^3}\right]_0^a = \dfrac{a^3b}{6}.$

Using the parallel axis theorem, we can calculate the moment of inertia with respect to the axis at the centroid, I_{cx}. Or $I_x = I_{cx} + Ad^2$, where d is the normal distance between axis x and that parallel passing through the centroid, or $\bar{y}_c$. Therefore

$I_{cx} = I_x - A\bar{y}_c^2 = \dfrac{81ab^3}{140} - \dfrac{3ab}{4}\left(\dfrac{3b}{7}\right)^2 = \dfrac{108}{245}ab^3.$ Similarly, $I_{cy} = I_y -$

$A\bar{x}_c^2 = \dfrac{a^3b}{6} - \dfrac{3ab}{4}\left(\dfrac{2a}{5}\right)^2 = \dfrac{7}{150}a^3b.$

<u>Polar moment of inertia</u> is the second moment of area with reference to a desired point. Considering the origin,

$$J_o = \int r^2 dA = \int (x^2 + y^2) dA = I_x + I_y = \frac{81ab^3}{140} + \frac{a^3b}{6} = \frac{ab}{420}(70a^2 + 243b^2).$$

we can calculate the polar moment of inertia with respect to the centroid, $J_c = I_{cx} + I_{cy}.$ Or $J_c = \frac{108ab^3}{245} + \frac{7a^3b}{150} = \frac{ab}{7350}(343a^2 + 3240b^2)$

Table 6 lists the results for a cubic polynomial shape.

Table 6 Results for cubic polynomial shape

Shape	Area	Centroid, $\bar{y}_c$	I_x	I_{cx}	J_o	J_c
Cubic	$\dfrac{ab}{4}$	$\dfrac{3b}{7}$	$\dfrac{81ab^3}{140}$	$\dfrac{108}{245}ab^3$	$\dfrac{ab}{420}(70a^2 + 243b^2)$	$\dfrac{7a^3b}{150} = \dfrac{ab}{7350}(343a^2 + 3240b^2)$

6 n-DEGREE POLYNOMIAL SHAPE-SPANDREL

As shown in Figure 6, consider a polynomial of degree n with its vertex at the origin. The polynomial equation reads $y = b\left(\dfrac{x}{a}\right)^{n}$. The area under the polynomial, Spandrel is of interest in engineering. A differential element $dA = dxdy$ is used for the following calculation.

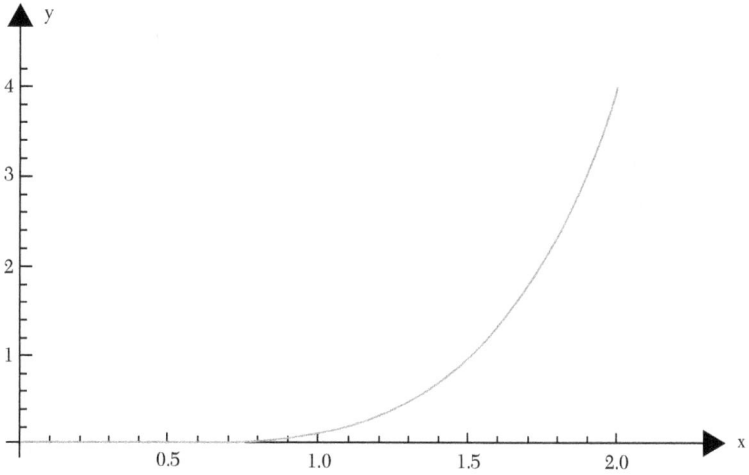

Figure 6 A spandrel area shape with a=2 and b=4

<u>Area,</u> we calculate the area using integration of the area differential element $dA = dxdy$. Using the equation of polynomial, we have

$$A = \int\limits_{0}^{a}\int\limits_{0}^{y} dxdy = \int\limits_{0}^{a} dx \int\limits_{0}^{y} dy = \int\limits_{0}^{a} y dx = \int\limits_{0}^{a} b\left(\frac{x}{a}\right)^{n} dx = \frac{b}{a^{n}}\left[\frac{x^{n+1}}{n+1}\right]_{0}^{a} = \frac{ab}{n+1}.$$

Simply, subtracting this area from that of the enclosing rectangle, hb we get the value of the area over the polynomial, or $\dfrac{nab}{n+1}$. Note that for $n = 2$, previous results for a parabolic are obtained.

<u>Centroid</u> is the first moment of area scaled by total area. Or

$$\bar{y}_{c} = \frac{\int y dA}{\int dA} = \frac{\int y dA}{A}. \quad \text{But} \quad \int y dA = \int\limits_{0}^{a}\int\limits_{0}^{y} y dxdy = \int\limits_{0}^{a} dx \int\limits_{0}^{y} y dy = \int\limits_{0}^{a}\left(\frac{y^{2}}{2}\right) dx.$$

After substituting for y and performing the integration, gives,

$$\int_0^a \left(\frac{y^2}{2}\right)dx = \frac{b^2}{2a^{2n}}\int_0^a x^{2n}dx = \frac{b^2}{2a^{2n}}\left[\frac{x^{2n+1}}{2n+1}\right]_0^a = \frac{ab^2}{2(2n+1)}.$$ Therefore,

$$\bar{y}_c = \frac{\dfrac{ab^2}{2(2n+1)}}{\dfrac{ab}{n+1}} = \frac{(n+1)b}{4n+2}.$$

Similarly, we can calculate the x-coordinate of the centroid,

$$\bar{x}_c = \frac{\int x dA}{\int dA} = \frac{\int x dA}{A}.$$ But $\int_0^a x dA - \int_0^a\int_0^y x dx dy = \int_0^a x dx\int_0^y dy = \int_0^a yx dx.$

After substituting for y and performing the integration,

gives, $\int_0^a yx dx = \frac{b}{a^n}\int_0^a x^{n+1}dx = \frac{b}{a^n}\left[\frac{x^{n+2}}{n+2}\right]_0^a = \frac{a^2b}{n+2}.$ Therefore,

$$\bar{x}_c = \frac{\dfrac{a^2b}{n+2}}{\dfrac{ab}{n+1}} = \frac{(n+1)a}{n+2}.$$

<u>Moment of inertia</u> is the second moment of area with reference to a desired axis. Considering x-axis as reference,

$$I_x = \int y^2 dA = \int_0^a dx\int_0^y y^2 dy = \int_0^a \left(\frac{y^3}{3}\right)dx = \frac{b^3}{3}\int_0^a\left(\frac{x^{3n}}{a^{3n}}\right)dx = \frac{b^3}{3a^{3n}}\left(\frac{a^{3n+1}}{3n+1}\right).$$

Performing the integrations gives $I_x = \dfrac{ab^3}{9n+3}.$ Similarly, we

can calculate $I_y = \int x^2 dA = \int_0^a x^2 dx\int_0^y dy = \int_0^a x^2 y dx = b\int_0^a\left(\frac{x^{n+2}}{a^n}\right)dx = \frac{b}{a^n}$

$\left(\dfrac{a^{n+3}}{n+3}\right) = \dfrac{a^3b}{n+3}.$

Using the parallel axis theorem, we can calculate the moment of inertia with respect to the axis at the centroid, I_{cx}. Or $I_x = I_{cx} + Ad^2$, where d is the normal distance between axis x and that parallel passing through the centroid, or $\bar{y}_c$. Therefore

$$I_{cx} = I_x - A\bar{y}_c^2 = \frac{ab^3}{9n+3} - \frac{ab}{n+1}\left[\frac{(n+1)b}{4n+2}\right]^2 = \left[\frac{7n^2+4n+1}{12(3n+1)(2n+1)^2}\right]ab^3.$$

Similarly, $I_{cy} = I_y - A\bar{x}_c^2 = \dfrac{a^3 b}{n+3} - \dfrac{ab}{n+1}\left[\dfrac{(n+1)a}{n+2}\right]^2 =$

$\left[\dfrac{1}{(n+3)(n+2)^2}\right]a^3 b.$

Polar moment of inertia is the second moment of area with reference to a desired point. Considering the origin,

$J_o = \int r^2 dA = \int \left(x^2 + y^2\right) dA = I_x + I_y = \dfrac{ab^3}{9n+3} + \dfrac{a^3 b}{n+3} = \dfrac{ab}{2}\left(\dfrac{a^2}{n+3} + \dfrac{b^2}{9n+3}\right).$

We can also calculate the polar moment of inertia with respect to the centroid, J_c. Or $J_c = I_{cx} + I_{cy}.$ Therefore

$J_c = \left[\dfrac{7n^2 + 4n + 1}{12(3n+1)(2n+1)^2}\right]ab^3 + \left[\dfrac{1}{(n+3)(n+2)^2}\right]a^3 b = ab$

$\left[\dfrac{a^2}{(n+3)(n+2)^2} + \dfrac{(7n^2 + 4n + 1)b^2}{12(3n+1)(2n+1)^2}\right].$

Table 7 lists the results for a parabolic shape.

TABLE 7 Results for Spandrel

Shape	Area	Centroid, yc	Ix	Icx	Jo	Jc
Spandrel, order n	$\dfrac{ab}{n+1}$	$\dfrac{(n+1)b}{4n+2}$	$\dfrac{b^3}{3a^{3n}}\left(\dfrac{a^{3n+1}}{3n+1}\right)$	$\left[\dfrac{7n^2 + 4n + 1}{12(3n+1)(2n+1)^2}\right]ab^3$	$\dfrac{ab}{2}\left(\dfrac{a^2}{n+3} + \dfrac{b^2}{9n+3}\right)$	$ab\left[\dfrac{a^2}{(n+3)(n+2)^2} + \dfrac{(7n^2 + 4n + 1)b^2}{12(3n+1)(2n+1)^2}\right]$

These results are comparable to those obtained in the previous sections for two- and three-degree polynomials.

7   SINUSOIDAL SHAPES

Consider a sinusoidal shape, $y = b\sin\left(\dfrac{\pi x}{a}\right)$ which is bounded between coordinate axes in the first quadrant (i.e., $x > 0$ and $y > 0$) as shown in Figure 7, for example for $a = 2$ and $b = 3$.

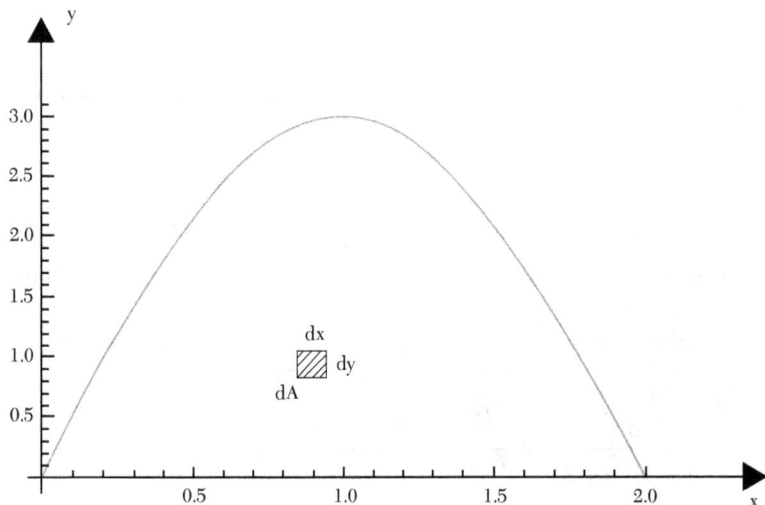

Figure 7 A sinusoidal area shape, $3\sin\left(\dfrac{\pi x}{2}\right)$.

Area, we calculate the area using integration of the differential area element $dA = dxdy$. Using the equation of the profile, we have $A = \int dxdy = \int_0^y dy \int_0^a dx = \int_0^a ydx = b\int_0^a \sin\left(\dfrac{\pi x}{a}\right)dx$. Performing the integration gives $A = b\left[-\dfrac{a}{\pi}\cos\dfrac{\pi x}{a}\right]_0^a = \dfrac{2ab}{\pi}$. Simply, subtracting this area from that of the enclosing rectangle, ab we get the value of the area over the sinusoidal curve, or $ab\left(1 - \dfrac{2}{\pi}\right) \cong 0.363ab$.

<u>Centroid</u> is the first moment of area scaled by total area. The y-coordinate of the centroid is $\bar{y}_c = \dfrac{\int y\, dA}{\int dA} = \dfrac{\int y\, dA}{A}$.

But $\int y\, dA = \displaystyle\int_0^a\int_0^y y\, dx\, dy = \int_0^a dx \int_0^y y\, dy = \int_0^a \left(\dfrac{y^2}{2}\right) dx$. After substituting

for y and performing the integration, we get

$$\int_0^a \left(\dfrac{y^2}{2}\right) dx = \dfrac{b^2}{2}\int_0^a \sin^2\left(\dfrac{\pi x}{a}\right) dx = \dfrac{b^2}{4}\int_0^a \left(1 - \cos\dfrac{2\pi x}{a}\right) dx = \dfrac{b^2}{4}\left[x - \dfrac{a}{2\pi}\sin\right.$$

$$\left.\dfrac{2\pi x}{a}\right]_0^a = \dfrac{ab^2}{4}. \text{ Therefore, } \bar{y}_c = \dfrac{\dfrac{ab^2}{4}}{2ab} = \dfrac{\pi b}{8}.$$

Similarly, we can calculate the x-coordinate of the centroid,

$\bar{x}_c = \dfrac{\int x\, dA}{\int dA} = \dfrac{\int x\, dA}{A}$. But $\int x\, dA = \displaystyle\int_0^a\int_0^y x\, dx\, dy = \int_0^a x\, dx\int_0^y dy = \int_0^a yx\, dx$. After

substituting for y and performing the integration, we get

$$\int_0^a yx\, dx = b\int_0^a x\sin\left(\dfrac{\pi x}{a}\right) dx = \left[-\dfrac{ab}{\pi}x\cos\dfrac{\pi x}{a}\right]_0^a + \underbrace{\dfrac{ab}{\pi}\int_0^a \cos\dfrac{\pi x}{a}\, dx}_{=0} = \dfrac{a^2 b}{\pi}.$$

Therefore, $\bar{x}_c = \dfrac{\dfrac{a^2 b}{\pi}}{\dfrac{2ab}{\pi}} = \dfrac{a}{2}$. This result confirms that from the shape

symmetry (i.e., the shape is symmetric about the line at $= \dfrac{a}{2}$).

<u>Moment of inertia</u> is the second moment of area with reference to a desired axis. Considering x-axis as reference,

$$I_x = \int y^2\, dA = \int_0^a dx\int_0^y y^2\, dy = \int_0^a \left(\dfrac{y^3}{3}\right) dx = \dfrac{b^3}{3}\int_0^a \sin^3\left(\dfrac{\pi x}{a}\right) dx.$$

Rewrite the latter integral as $\dfrac{b^3}{3}\int_0^a \sin\dfrac{\pi x}{a}\sin^2\left(\dfrac{\pi x}{a}\right) dx = \dfrac{b^3}{3}\int_0^a \sin\dfrac{\pi x}{a}$

$$\left(1-\cos^2\left(\frac{\pi x}{a}\right)\right) dx$$ and performing the integration, we get

$$\frac{b^3}{3}\left[-\frac{a}{\pi}\cos\frac{\pi x}{a}\right]_0^a - \frac{b^3}{3}$$

$$\int_0^a \sin\frac{\pi x}{a}\cos^2\left(\frac{\pi x}{a}\right) dx = \frac{2ab^3}{3\pi} - \frac{b^3}{3}\left[-\frac{a}{3\pi}\cos^3\left(\frac{\pi x}{a}\right)\right]_0^a = \frac{2ab^3}{3\pi} - \frac{2ab^3}{9\pi}$$

$$= \frac{4ab^3}{9\pi} \cong 0.1415a^3 b.$$

Similarly, we can calculate $I_y = \int x^2 dA = \int_0^y dy \int_0^a x^2 dx = \int_0^a yx^2 dx =$

$$b\int_0^a x^2 \sin\left(\frac{\pi x}{a}\right) dx.$$ Performing the integration gives

$$I_y = \left[-\frac{ab}{\pi}x^2\cos\frac{\pi x}{a}\right]_0^a + \frac{2ab}{\pi}\int_0^a x\cos\frac{\pi x}{a} dx = \frac{a^3 b}{\pi} + \frac{2ab}{\pi}\int_0^a x\cos\frac{\pi x}{a} dx.$$

The latter integral reads $\int_0^a x\cos\frac{\pi x}{a} dx = \underbrace{\left[x\frac{a}{\pi}\sin\frac{\pi x}{a}\right]_0^a}_{=0} - \frac{a}{\pi}\int_0^a$

$$\sin\frac{\pi x}{a} dx = \left[\frac{a^2}{\pi^2}\cos\frac{\pi x}{a}\right]_0^a = -\frac{2a^2}{\pi^2}.$$ Therefore, $I_y = \frac{a^3 b}{\pi} - \frac{2ab}{\pi}\cdot\frac{2a^2}{\pi^2} =$

$$\frac{\left(\pi^2-4\right)a^3 b}{\pi^3} \cong 0.1893a^3 b.$$

Using the parallel axis theorem, we can calculate the moment of inertia with respect to the axis at the centroid parallel to the x-axis, I_{cx}. Or $I_x = I_{cx} + Ad^2$, where d is the normal distance between axis x and that parallel passing through the centroid, or $\bar{y}_c$. Therefore

$$I_{cx} = I_x - A\bar{y}_c^2 = \frac{4ab^3}{9\pi} - \left(\frac{2ab}{\pi}\right)\left(\frac{\pi b}{8}\right)^2 = \frac{\left(128-9\pi^2\right)ab^3}{288\pi}.$$ Similarly,

$$I_{cy} = I_y - A\bar{x}_c^2 = \frac{\left(\pi^2-4\right)a^3 b}{\pi^3} - \left(\frac{2ab}{\pi}\right)\left(\frac{a}{2}\right)^2 = \frac{\left(\pi^2-8\right)a^3 b}{2\pi^3} \cong 0.03a^3 b.$$

<u>Polar moment of inertia</u> is the second moment of area with reference to a desired point. Considering the origin,

$$J_o = \int r^2 dA = \int \left(x^2 + y^2\right) dA = I_x + I_y = \frac{4ab^3}{9\pi} + \frac{\left(\pi^2 - 4\right)a^3 b}{\pi^3} =$$

$$\frac{\left(13\pi^2 - 36\right)a^3 b}{9\pi^3} \cong 0.3308 a^3 b.$$

We can also calculate the polar moment of inertia with respect to the centroid, J_c. Or $J_c = I_{cx} + I_{cy}$. Therefore

$$J_c = \frac{\left(128 - 9\pi^2\right)ab^3}{288\pi} + \frac{\left(\pi^2 - 8\right)a^3 b}{2\pi^3} = \frac{\left(272\pi^2 - 9\pi^4 - 1152\right)a^3 b}{288\pi^3} \cong$$

$0.0734 a^3 b.$

Table 8 lists the results for the sinusoidal shape area:

TABLE 8 Some results for sinusoidal shape

Shape	Area	Centroid, yc	Ix	Icx	Jo	Jc
sinusoidal	$\dfrac{2ab}{\pi}$	$\dfrac{\pi b}{8}$	$\dfrac{4ab^3}{9\pi}$	$\dfrac{\left(128 - 9\pi^2\right)ab^3}{288\pi}$	$\dfrac{\left(13\pi^2 - 36\right)a^3 b}{9\pi^3}$	$\dfrac{\left(272\pi^2 - 9\pi^4 - 1152\right)a^3 b}{288\pi^3}$

8   TRIANGULAR SHAPES

We consider an equilateral triangle shape with side a. Its area, centroid, and moment of inertia are calculated using related integrals.

A triangular shape can be the cross-section shape of beams, rods, or hydraulic gates for example.

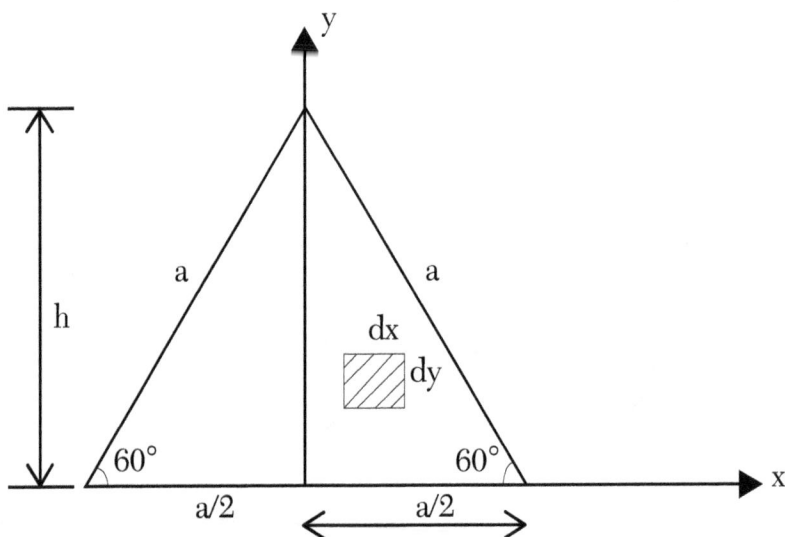

Figure 8 A triangular area shape

<u>Area</u> is equal to $\dfrac{a^2\sqrt{3}}{4}$, from geometry. We calculate the area using the integration of the area differential element $dA = dxdy$ as shown in Figure 8. Using the geometrical properties of the triangle we have $y = h\left(1 - \dfrac{2x}{a}\right)$, where $h = \dfrac{a\sqrt{3}}{2}$ is the height of the triangle.

Therefore, we have $A = \displaystyle\int\int dxdy = \int\limits_0^y dy \int\limits_{-a/2}^{a/2} dx = 2h \int\limits_0^{a/2}\left(1 - \dfrac{2x}{a}\right)dx =$

$a\sqrt{3}\left[x - \dfrac{x^2}{a}\right]_0^{a/2} = \dfrac{a^2\sqrt{3}}{4}.$

<u>Centroid</u> is the first moment of area divided by its total area and is located at $(0, \bar{y})$ when symmetry about y-axis exists. Or $\bar{y} = \dfrac{\int y dA}{\int dA} =$

$\dfrac{\int y dA}{A}$. But $\int y dA = 2 \int_0^{y} y dy \int_0^{a/2} dx = \int_0^{a/2} y^2 dx = h^2 \int_0^{a/2} \left(1 - \dfrac{2x}{a}\right)^2 dx = \dfrac{a^3}{8}$.

Therefore, $\bar{y} = \dfrac{a^3/8}{a^2 \sqrt{3}/4} = \dfrac{a\sqrt{3}}{6}$.

<u>Moment of inertia</u> is the second moment of area with reference to a desired axis. Considering x-axis as reference,

$$I_x = \int y^2 dA = 2 \int_0^{y} y^2 dy \int_0^{a/2} dx = \dfrac{2}{3} \int_0^{a/2} y^3 dx = \dfrac{2h^3}{3} \int_0^{a/2} \left(1 - \dfrac{2x}{a}\right)^3 dx.$$

Performing the integral gives

$$I_x = \dfrac{2h^3}{3} \int_0^{a/2} \left(1 - \dfrac{2x}{a}\right)^3 dx = \dfrac{2h^3}{3} \left[x - \dfrac{3}{a} x^2 + \dfrac{4}{a^2} x^3 - \dfrac{2}{a^3} x^4 \right]_0^{a/2} = \dfrac{\sqrt{3}}{32} a^4.$$

Similarly, we can calculate the moment of inertia with respect to the y axis. Or $I_y = \int x^2 dA = 2 \int_0^{y} dy \int_0^{a/2} x^2 dx = 2 \int_0^{a/2} yx^2 dx = 2h \int_0^{a/2} x^2 \left(1 - \dfrac{2x}{a}\right) dx.$

Performing the integral gives

$$I_y = 2h \int_0^{a/2} x^2 \left(1 - \dfrac{2x}{a}\right) dx = 2h \left[\dfrac{x^3}{3} - \dfrac{x^4}{2a} \right]_0^{a/2} = \dfrac{\sqrt{3}}{96} a^4.$$

Using the parallel axis theorem, we can calculate the moment of inertia with respect to the parallel axis at the centroid, I_{cx}. Or $I_x = I_{cx} + Ad^2$, where d is the normal distance between axis x and the parallel axis passing through the centroid, or $\bar{y}$. Therefore

$$I_{cx} = I_x - Ad^2 = \dfrac{\sqrt{3}}{32} a^4 - \dfrac{a^2 \sqrt{3}}{4} \left(\dfrac{a\sqrt{3}}{6}\right)^2 = \dfrac{\sqrt{3}}{96} a^4.$$

<u>Polar moment of inertia</u> is the second moment of area with reference to a desired point. Considering the origin as the reference,

$$J_o = \int r^2 dA = \int (x^2 + y^2) dA = I_x + I_y = \dfrac{\sqrt{3}}{32} a^4 + \dfrac{\sqrt{3}}{96} a^4 = \dfrac{\sqrt{3}}{24} a^4.$$

9   RECTANGULAR SHAPES

We consider a rectangular shape with base b and height h. The area, centroid, and moment of inertia are calculated using related integrals.

A rectangular shape can be the cross-section shape of a beam, rods, or hydraulic gates for example.

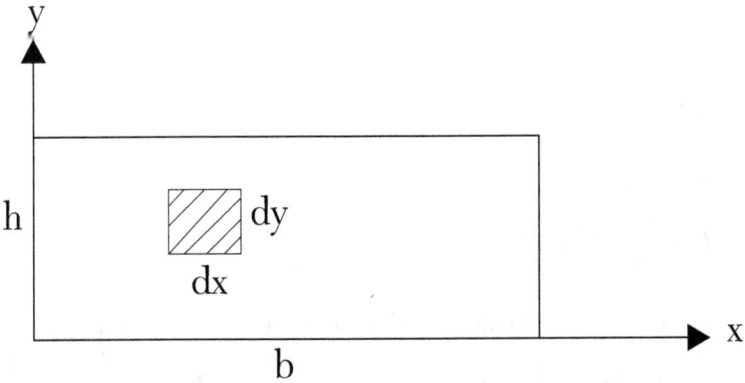

Figure 9 A rectangular area shape

<u>Area</u> is equal to bh, from geometry. We can calculate the area using integration of the differential area element $dA = dxdy$ as shown in Figure 9. Therefore, considering a coordinate system, $x - y$ with its origin at the down left vertex, we have

$$A = \int\int dxdy = \int_0^h dy \int_0^b dx = h\int_0^b dx = bh.$$

<u>Centroid</u> is the first moment of area divided by total area. We can write $\bar{y} = \dfrac{\int ydA}{A} = \dfrac{\int_0^h ydy \int_0^b dx}{bh}$. Therefore. After performing the integration operation, we get $\bar{y} = \dfrac{\int_0^h ydy \int_0^b dx}{bh} = \dfrac{b\left[y^2/2\right]_0^h}{bh} = \dfrac{h}{2}$.

Similarly, $\bar{x} = \dfrac{\int xdA}{A} = \dfrac{\int_0^b xdx \int_0^h dy}{bh} = \dfrac{h\left[x^2/2\right]_0^b}{bh} = \dfrac{b}{2}$. Therefore, the centroid is at $C(b/2, h/2)$. Theses results are consistent with those obtained from the symmetry property of the rectangular shape.

<u>Moment of inertia</u> is the second moment of area with reference to a desired axis. Considering x-axis as the reference, we can write $I_x = \int y^2 dA = \int\limits_0^h y^2 dy \int\limits_0^b dx$. Therefore, $I_x = b \int\limits_0^h y^2 dy = b\left[y^3 / 3 \right]_0^h = \dfrac{bh^3}{3}$. Similarly, moment of inertia about y-axis reads $I_y = \int x^2 dA = \int\limits_0^b x^2 dx \int\limits_0^h dy = \dfrac{hb^3}{3}$.

Using parallel axis theorem, we can calculate the moment inertia about a system of coordinates, $x_c - y_c$ with its origin at the centroid. Or $I_{xc} = I_x - A\bar{y}^2 = \dfrac{bh^3}{3} - bh\left(\dfrac{h}{2}\right)^2 = \dfrac{bh^3}{12}$, and

$$I_{yc} = I_y - Ad^2 = \dfrac{hb^3}{3} - bh\left(\dfrac{b}{2}\right)^2 = \dfrac{hb^3}{12}.$$

<u>Polar moment of inertia</u> about origin O, is $J_o = I_x + I_y = \dfrac{bh}{3}\left(h^2 + b^2\right)$. Also, the polar moment of inertia about centroid reads

$$J_c = I_{xc} + I_{yc} = \dfrac{bh}{12}\left(h^2 + b^2\right).$$

We can also calculate J_c using the parallel axis theorem, as

$$J_c = J_o - Ar^2 = \dfrac{bh}{3}\left(h^2 + b^2\right) - bh\left(\dfrac{b^2}{4} + \dfrac{h^2}{4}\right) = \dfrac{bh}{12}\left(h^2 + b^2\right).$$

10   COMPLEX SHAPES

A complex shape can be divided into simple shapes. This facilitates the calculation of the area properties for complex shapes using those of simple shape components. The only requirement is this that for each simple shape property, for example moment of inertia, we consider a common reference axis. Again, parallel axis theorem can be used for transformation of properties.

10.1 Example: A semi-circle with a semi-elliptical hole

We consider a semi-circular shape with radius R and with its center located at the origin. A semi elliptical shape is taken away from the original area with its semi radii being $a = \alpha R$ and $b = \beta R$, respectively ($0 < \alpha < 1$ and $0 < \beta < 1$, are constants). Figure 10 shows the complex shape.

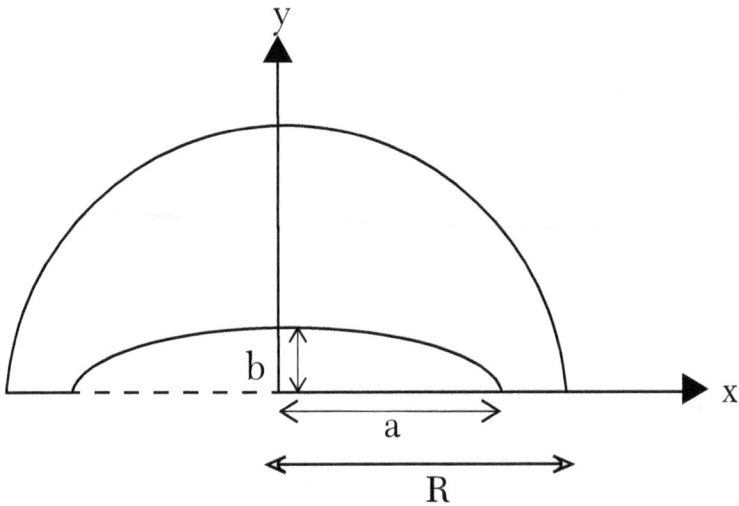

Figure 10 A complex semi circular area shape with a semi ellipse hole

Area can be obtained by subtracting the area of the semi-ellipse from that of the semi-circle. As shown in the previous sections (see Table 3 and Table 5), we have $A = \dfrac{\pi R^2}{2} - \dfrac{\pi \alpha \beta R^2}{2} = \dfrac{\pi(1-\alpha\beta)}{2} R^2.$

<u>Centroid</u> is located at $\bar{y}$ distanced from the x- axis on the y-axis. Using the obtained results from the previous sections, we have

$$\bar{y} = \frac{\left(\dfrac{\pi R^2}{2}\right)\left(\dfrac{4R}{3\pi}\right) - \left(\dfrac{\pi\alpha\beta R^2}{2}\right)\left(\dfrac{4\beta R}{3\pi}\right)}{\dfrac{\pi(1-\alpha\beta)}{2}R^2} = \frac{4(1-\beta^2)}{3\pi(1-\alpha\beta)}R.$$

<u>Moment of inertia</u> with respect to x-axis is simply that of semi-ellipse part subtracted from the semi-circle's. Therefore, $I_x = \dfrac{\pi R^4}{8} - \dfrac{\pi\alpha\beta^3 R^4}{8} = \dfrac{\pi R^4}{8}(1-\alpha\beta^3)$. But moment of inertia with respect to the axis at the centroid of the complex shape requires the application of the parallel axis. For the semi-circle part we have, using the previously obtained results, $I_{cx-circle} = \left(\dfrac{9\pi^2 - 64}{72\pi}\right)R^4 + \left(\dfrac{\pi R^2}{2}\right)\left(\dfrac{4R}{3\pi} - \dfrac{4(1-\beta^2)R}{3\pi(1-\alpha\beta)}\right)^2$. For the semi-ellipse part we have, using the previously obtained results,

$$I_{cx-ellipse} = \left(\frac{9\pi^2 - 64}{72\pi}\right)\alpha\beta^3 R^4 + \left(\frac{\pi\alpha\beta R^2}{2}\right)\left(\frac{4\beta R}{3\pi} - \frac{4(1-\beta^2)R}{3\pi(1-\alpha\beta)}\right)^2.$$

As a numerical example, let $\alpha = 0.75$, and $\beta = 0.5$. Therefore, we get the following results as shown in Table 9.

TABLE 9 Results for a composite shape

Shape	Area	Centroid, y	Ix	Icx_circle	Icx_ellipse
composite	$0.9817R^2$	$0.5093R$	$0.3559R^4$	$0.39795R^4$	$0.06228R^4$

10.2 Example: A rectangle with circular segment sides

We consider a rectangular shape with base b and height h. The two vertical sides are composed of two circular segments where the corresponding circles' centers are located at $h/2$ with radii R. Figure 11 shows the complex shape.

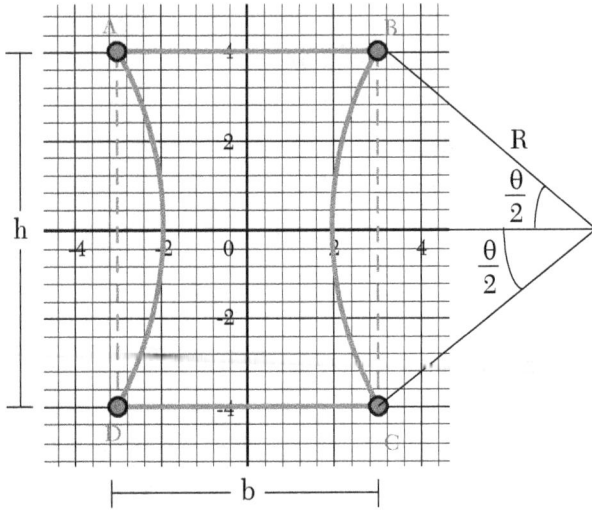

Figure 11 A complex rectangular area shape with circular segment sides

Considering the complex shape composed of two parts: a rectangle and two circular segments. We use the results obtained in previous sections for calculations.

Area can be obtained by subtracting the area of the segments from that of the rectangle. As shown in previous sections we have $A = bh - 2\dfrac{R^2}{2}(\theta - \sin\theta)$. But, from geometry, $h = 2R\sin\dfrac{\theta}{2}$. Therefore, we have $A = 2bR\sin\dfrac{\theta}{2} - R^2(\theta - \sin\theta)$. Note that θ is the angle corresponding to the circular arc.

Centroid is simply at the geometrical center of the shape, due to symmetry. Therefore, $\bar{y} = \dfrac{h}{2} = R\sin\dfrac{\theta}{2}$.

Moment of inertia with respect to the axis at the centroid and parallel to the base is simply that of rectangle part subtracted by the circular segments. Therefore, $I_{cx} = \dfrac{bh^3}{12} - 2\dfrac{R^4}{24}\left(3\theta - 3\sin\theta - 2\sin\theta\sin^2\dfrac{\theta}{2}\right)$. Therefore, after simplifications, we have $I_{cx} = \dfrac{R^3}{12}\left[8b\sin^3(\theta/2) - R\left(3\theta - 3\sin\theta - 2\sin\theta\sin^2\dfrac{\theta}{2}\right)\right]$. Similarly, the moment of inertia with respect to axis at the centroid and parallel to the sides is that of the rectangle part subtracted by the circular segments. But the distance from the centroid of the circular arc to the y-axis reads, from geometry, $d = \dfrac{b}{2} + R\cos\dfrac{\theta}{2} - \dfrac{4R\sin^3\left(\dfrac{\theta}{2}\right)}{3(\theta - \sin\theta)}.$

Therefore, the moment of inertia of the circular sectors with respect to the y-axis , using the parallel axis theorem, reads

$$I_{ys} = \frac{R^4}{8}\left[\theta - \sin\theta + 2\sin\theta\sin^2\left(\frac{\theta}{2}\right) - \frac{64}{9}\cdot\frac{\sin^6\left(\frac{\theta}{2}\right)}{\theta - \sin\theta}\right] + \frac{R^2}{2}(\theta - \sin\theta)$$

$$\left[\frac{b}{2} + R\cos\frac{\theta}{2} - \frac{4R\sin^3\left(\frac{\theta}{2}\right)}{3(\theta - \sin\theta)}\right]^2.$$

Therefore, for the complex shape we can write

$$I_{cy} = \frac{hb^3}{12} - 2I_{ys} = \frac{b^3 R\sin\left(\frac{\theta}{2}\right)}{6} - \frac{R^4}{4}\left[\theta - \sin\theta + 2\sin\theta\sin^2\left(\frac{\theta}{2}\right) - \frac{64}{9}\right.$$

$$\left.\frac{\sin^6\left(\frac{\theta}{2}\right)}{\theta - \sin\theta} - R^2(\theta - \sin\theta)\left[\frac{b}{2} + R\cos\frac{\theta}{2} - \frac{4R\sin^3\left(\frac{\theta}{2}\right)}{3(\theta - \sin\theta)}\right]^2.$$

From these results, the polar moment of inertia about centroid can be calculated $J_c = I_{cx} + I_{cy}.$

10.3 Example: A semi-circle with a triangular shape hole

We consider a semi-circle with radius R when a triangular shape with base b and height h. Is subtracted from it, as show in Figure 12. The complex shape is symmetric about the y-axis with conditions that $\dfrac{b}{2} < R$ and $h < R.$

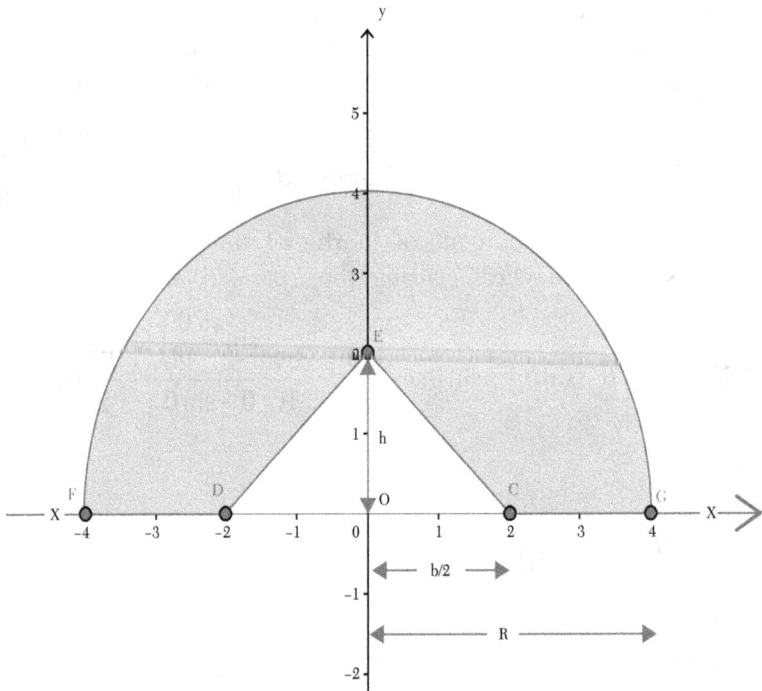

Figure 12 A complex semi-circular area shape with a triangular shape hole

Considering the complex shape composed of two parts: a rectangle and two circular segments. We use the results obtained in the previous sections for calculations.

Area can be obtained by subtracting the area of the triangle from that of the semi-circle. Or $A = \dfrac{\pi R^2 - bh}{2}$.

Centroid is simply at the geometrical center of the shape. Therefore, with reference to the base $x-x$ axis we have $\bar{y} = \dfrac{\left(\pi R^2 / 2\right)\left(4R / 3\pi\right) - \left(bh / 2\right)\left(h / 3\right)}{\left(\pi R^2 - bh\right) / 2}$. After some simplifica

tions, we have $\bar{y} = \dfrac{4R^3 - bh^2}{3\left(\pi R^2 - bh\right)}$. Due to symmetry, $\bar{x} = 0$.

<u>Moment of inertia</u> with respect to the x-x axis can be calculated as that of triangle, $\dfrac{bh^3}{12}$ from the semi-circle's, $\dfrac{\pi R^4}{8}$.

Or $I_x = \dfrac{\pi R^4}{8} - \dfrac{bh^3}{12}$. After using the parallel axis theorem, we can calculate the moment inertia about the centroid as $I_{xc} = \left(\dfrac{\pi R^4}{8} - \dfrac{bh^3}{12} \right) - \left(\dfrac{\pi R^2 - bh}{2} \right) \left[\dfrac{4R^3 - bh^2}{3(\pi R^2 - bh)} \right]^2$.

Or, after simplifying this expression we get $I_{cx} =$

$$\frac{(64 - 9\pi^2)R^6 + 9\pi bh R^4 - 32bh^2 R^3 + 6\pi bh^3 R^2 - 2b^2 h^4}{72(bh - \pi R^2)}.$$

Similarly, using symmetry, the moment of inertia with respect to the y-axis can be written as $I_y = I_{cy} = \dfrac{\pi R^4}{8} - \dfrac{hb^3}{12}$.

From these results, the polar moment of inertia about the centroid can be calculated $J_c = I_{cx} + I_{cy}$.

11 A CANTILEVER BEAM WITH CUBIC LOAD DISTRIBUTION

A cantilever beam is loaded with a cubically distributed load, as shown in Figure 13. The load is given as $q = \omega \left(\dfrac{x}{L} \right)^3$ where, ω is the maximum load density per unit length (in N/m), and L beam length. We calculate the equivalent load, its acting location x_c, and the distribution of the shear and bending moment for this beam.

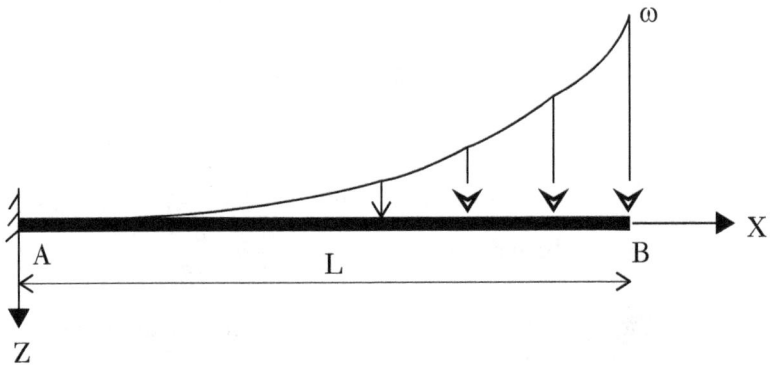

Figure 13 A Cantilever beam with cubic load distribution.

Equivalent load W, is the area under the distribution. Therefore, we can write $W = \int_0^L q\,dx = \int_0^L \omega \left(\dfrac{x}{L} \right)^3 dx$. Performing the integration

gives $\int_0^L \omega \left(\dfrac{x}{L} \right)^3 dx = \omega \left[\dfrac{L^4}{4L^3} \right] = \dfrac{\omega L}{4}$.

Centroid of the load, x_c is where W is acting. Therefore,

$$x_c = \frac{\int_0^L xq\,dx}{\dfrac{\omega L}{4}} = \frac{\int_0^L x\omega \left(\dfrac{x}{L} \right)^3}{\dfrac{\omega L}{4}} = \frac{\dfrac{\omega}{L^3} \left(\dfrac{L^5}{5} \right)}{\dfrac{\omega L}{4}} = \frac{4L}{5}.$$ We can calcu-

late the support reaction force $R_A = \dfrac{\omega L}{4}$, and the moment

$M_A = -\left(\dfrac{\omega L}{4}\right)\left(\dfrac{4L}{5}\right) = -\dfrac{\omega L^2}{5}$, after writing the balance of forces in the z-direction and the that of the moments of point A.

Shear force distribution, V as a function of x, is given as $V = -\int q\,dx + C_1$ where, C_1 is determined by the boundary conditions for the shear force, for example $V|_{x=L} = 0$. Performing the integration, gives $V = -\int q\,dx + C_1 = -\dfrac{\omega}{4L^3}x^4 + C_1$. Applying the boundary conditions gives $-\dfrac{\omega}{4L^3}L^4 + C_1 = 0 \Rightarrow C_1 = \dfrac{\omega L}{4}$. Hence $V = -\dfrac{\omega}{4L^3}x^4 + \dfrac{\omega L}{4}$. Note that the shear force equation recovers the reaction force at the support A, or $V(x=0) = \dfrac{\omega L}{4}$.

Bending moment distribution, M as a function of x, is given as $M = \int V\,dx + C_2$ where, C_2 is determined by the boundary conditions for the moment, for example $M|_{x=L} = 0$. Performing the integration, gives $M = \int\left(-\dfrac{\omega}{4L^3}x^4 + \dfrac{\omega L}{4}\right)dx + C_2 = -\dfrac{\omega}{20L^3}x^5 + \dfrac{\omega L}{4}x + C_2$. Applying the boundary conditions gives $-\dfrac{\omega}{20L^3}L^5 + \dfrac{\omega L^2}{4} + C_2 = 0 \Rightarrow C_2 = -\dfrac{\omega L^2}{5}$. Hence $M = -\dfrac{\omega}{20L^3}x^5 + \dfrac{\omega L}{4}x - \dfrac{\omega L^2}{5}$. Note that the bending moment equation recovers the reaction moment at the support A, or $M(x=0) = -\dfrac{\omega L^2}{5}$.

12 A CANTILEVER BEAM WITH QUARTER-ELLIPSE LOAD DISTRIBUTION

A cantilever beam is loaded with an elliptical quarter distributed load, as shown in Figure 14. The load is given as $q = \omega\sqrt{1 - \left(\dfrac{x}{L}\right)^2}$ where, ω is the maximum load density per unit length (in N/m), and L beam length. We calculate the equivalent load, its acting location x_p, and the distribution of the shear and bending moment for this beam.

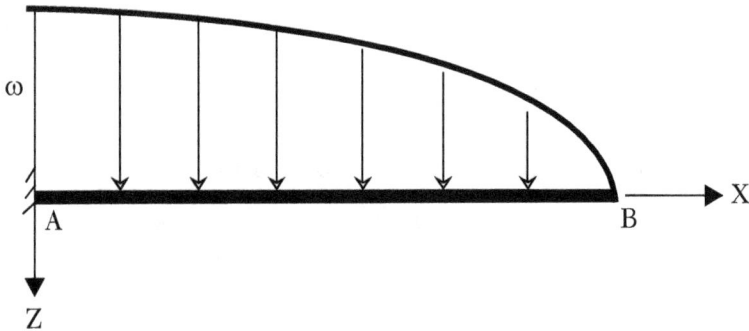

Figure 14 A Cantilever beam with quarter-ellipse load distribution.

Equivalent load W, is the area under the distribution. Therefore, we can write $W = \int_0^L q\,dx = \int_0^L \omega\sqrt{1 - \left(\dfrac{x}{L}\right)^2}\,dx.$

Let $\dfrac{x}{L} = \sin\alpha \Rightarrow dx = L\cos\alpha\,d\alpha$. Writing the integral in terms of the variable α and performing the integration gives

$$\omega\int_0^L \sqrt{1 - \left(\frac{x}{L}\right)^2}\,dx = \omega L\int_0^{\frac{\pi}{2}} \cos^2\alpha\,d\alpha = \frac{\omega L}{2}\int_0^{\frac{\pi}{2}}(1 + \cos 2\alpha)\,d\alpha = \frac{\omega L}{2}\left[\alpha + \frac{1}{2}\right.$$

$$\left. \sin 2\alpha\right]_0^{\frac{\pi}{2}} = \frac{\pi\omega L}{4}.$$

<u>Centroid of the load</u>, x_c is where W is acting. Therefore,

$$x_c = \frac{\omega \int_0^L x\sqrt{1-\left(\frac{x}{L}\right)^2}\,dx}{\dfrac{\pi\omega L}{4}} = \frac{4}{\pi L}\int_0^L x\sqrt{1-\left(\frac{x}{L}\right)^2}\,dx.$$ Writing the inte-

gral in terms of the variable α and performing the integration

gives $\displaystyle\int_0^L x\sqrt{1-\left(\frac{x}{L}\right)^2}\,dx = L^2\int_0^{\frac{\pi}{2}} \sin\alpha\cos^2\alpha\,d\alpha = -L^2\left(\frac{1}{3}\cos^3\alpha\right)\Big|_0^{\frac{\pi}{2}} = \frac{L^2}{3}.$

Therefore, $x_c = \left(\dfrac{4}{\pi L}\right)\left(\dfrac{L^2}{3}\right) = \dfrac{4L}{3\pi}$.

We can calculate the support reaction force $R_A = \dfrac{\pi\omega L}{4}$, and the

moment $M_A = -\left(\dfrac{\pi\omega L}{4}\right)\left(\dfrac{4L}{3\pi}\right) = -\dfrac{\omega L^2}{3}$, after writing the balance

of forces in the z-direction and the that of the moments about point A.

<u>Shear force distribution</u>, V as a function of x, is given as $V = -\int q\,dx + C_1$ where, C_1 is determined by boundary conditions for the shear force, for example $V|_{x=L} = 0$. Performing the inte-

gration, gives $V = -\int \omega\sqrt{1-\left(\frac{x}{L}\right)^2}\,dx + C_1$. Writing the integral

in terms of variable α (recall $\dfrac{x}{L} = \sin\alpha \Rightarrow dx = L\cos\alpha\,d\alpha$) gives

$-\dfrac{\omega L}{2}\int(1+\cos 2\alpha)\,d\alpha + C_1 = -\dfrac{\omega L}{2}\left(\alpha + \dfrac{1}{2}\sin 2\alpha\right) + C_1.$ Applying the

boundary condition, $V|_{x=L} = V|_{\alpha=\pi/2} = 0$ gives $-\dfrac{\pi\omega L}{4} + C_1 = 0 \Rightarrow$

$C_1 = \dfrac{\pi\omega L}{4}$. Hence, $V = -\dfrac{\omega L}{2}\left(\alpha + \dfrac{1}{2}\sin 2\alpha\right) + \dfrac{\pi\omega L}{4}$. Writing the

shear force equation in terms of the original variable x,

gives $V = -\dfrac{\omega L}{2}\left(\alpha + \dfrac{1}{2}\sin 2\alpha\right) + \dfrac{\pi\omega L}{4} = -\dfrac{\omega L}{2}(\alpha + \sin\alpha\cos\alpha) +$

$\dfrac{\pi\omega L}{4} = \dfrac{\omega L}{4}\left[\pi - \dfrac{2x}{L}\sqrt{1 - \dfrac{x^2}{L^2}} - 2\sin^{-1}\left(\dfrac{x}{L}\right)\right]$. Note that the shear

force equation recovers the reaction force at the support A, or

$V(x = 0) = \dfrac{\pi\omega L}{4}$.

Bending moment distribution, M as a function of x, is given as

$M = \int V\,dx + C_2$ where, C_2 is determined by boundary conditions

for the moment, for example $M\big|_{x=L} = 0$. writing the integral,

gives $M = \dfrac{\omega L}{4}\displaystyle\int\left(\pi - \dfrac{2x}{L}\sqrt{1 - \dfrac{x^2}{L^2}} - 2\sin^{-1}\dfrac{x}{L}\right)dx + C_2 = \dfrac{\pi\omega L}{4}x - \dfrac{\omega}{2}\displaystyle\int x$

$\sqrt{1 - \dfrac{x^2}{L^2}}dx - \dfrac{\omega L}{2}\displaystyle\int \sin^{-1}\dfrac{x}{L}dx + C_2$. But writing the former integral

in terms of the variable $\alpha\left(\text{recall,}\dfrac{x}{L} = \sin\alpha \Rightarrow dx = L\cos\alpha\,d\alpha\right)$

reads $-\dfrac{\omega}{2}\displaystyle\int x\sqrt{1 - \dfrac{x^2}{L^2}}dx = -\dfrac{\omega L^2}{2}\displaystyle\int \sin\alpha\cos^2\alpha\,d\alpha = \dfrac{\omega L^2}{6}\cos^3\alpha = \dfrac{\omega L^2}{6}$

$\left(1 - \dfrac{x^2}{L^2}\right)^{3/2}$. Similarly, the latter integral can be written as

$-\dfrac{\omega L}{2}\displaystyle\int \sin^{-1}\dfrac{x}{L}dx = -\dfrac{\omega L^2}{2}\displaystyle\int \alpha\cos\alpha\,d\alpha = -\dfrac{\omega L^2}{2}(\alpha\sin\alpha + \cos\alpha) = -$

$-\dfrac{\omega L^2}{2}\left[\dfrac{x}{L}\sin^{-1}\left(\dfrac{x}{L}\right) + \sqrt{1 - \dfrac{x^2}{L^2}}\right]$. After collecting all related

answers, we get the solution as $M = \dfrac{\pi\omega L}{4}x + \dfrac{\omega L^2}{6}\left(1 - \dfrac{x^2}{L^2}\right)^{3/2} -$

$\dfrac{\omega Lx}{2}\sin^{-1}\left(\dfrac{x}{L}\right) - \dfrac{\omega L^2}{2}\sqrt{1 - \dfrac{x^2}{L^2}} + C_2$. Applying the boundary con-

dition, $M\big|_{x=L} = 0$ gives $\dfrac{\pi\omega L^2}{4} - \dfrac{\pi\omega L^2}{4} + C_2 = 0 \Rightarrow C_2 = 0$. Hence, the bending moment equation, after simplification, reads

$$M = \frac{\omega L}{12}\left[3\pi x - 2L\left(2 + \frac{x^2}{L^2}\right)\sqrt{1 - \frac{x^2}{L^2}} - 6x\sin^{-1}\left(\frac{x}{L}\right) \right].$$ Note that the

bending moment equation recovers the reaction moment at the

support A, or $M(x=0) = -\dfrac{\omega L^2}{3}$.

13   A CANTILEVER BEAM WITH INVERSE COSINE LOAD DISTRIBUTION

A cantilever beam is loaded with an inverse cosine distributed load. The load is given as $q = \dfrac{2\omega}{\pi}\cos^{-1}\left(\dfrac{x}{L}\right)$ where, ω is the load density per unit length (e.g., N/m), and L beam length. In this section, we present the calculations for the equivalent load, its acting location x_c from the support, and the distribution of shear and bending moment along the length of this beam.

<u>Equivalent load</u> W, is the area under the distribution. Therefore, we can write $W = \displaystyle\int_0^L q\,dx = \dfrac{2\omega}{\pi}\int_0^L \cos^{-1}\left(\dfrac{x}{L}\right)dx$. Let $\cos^{-1}\left(\dfrac{x}{L}\right) = \alpha \Rightarrow \cos\alpha = \dfrac{x}{L}$ and $dx = -L\sin\alpha\,d\alpha$. Note that the limits of integral change to $\alpha = \dfrac{\pi}{2}$ for $x = 0$ and $\alpha = 0$ for $x = L$. Writing the integral in terms of the variable α and performing the integration gives

$$\dfrac{2\omega}{\pi}\int_0^L \cos^{-1}\left(\dfrac{x}{L}\right)dx = -\dfrac{2\omega L}{\pi}\int_{\frac{\pi}{2}}^0 \alpha\sin\alpha\,d\alpha = -\dfrac{2\omega L}{\pi}\left[-\alpha\cos\alpha + \sin\alpha\right]_{\frac{\pi}{2}}^0 =$$

$$\dfrac{2\omega L}{\pi}.$$

<u>Centroid of the load</u>, x_c is distance from the support where W is acting. Therefore, $x_c = \dfrac{\dfrac{2\omega}{\pi}\displaystyle\int_0^L x\cos^{-1}\left(\dfrac{x}{L}\right)dx}{\dfrac{2\omega L}{\pi}} = \dfrac{\displaystyle\int_0^L x\cos^{-1}\left(\dfrac{x}{L}\right)dx}{L}$. Writing the integral in terms of the variable α and performing the integration gives

$$\int_0^L x\cos^{-1}\left(\dfrac{x}{L}\right)dx = -L^2\int_0^{\frac{\pi}{2}}\alpha\sin\alpha\cos\alpha\,d\alpha = -L^2\left[\underbrace{\left(\alpha\dfrac{\sin^2\alpha}{2}\right)\Big|_{\frac{\pi}{2}}^0}_{=-\pi/4} - \right.$$

$$-\frac{1}{2}\int_{\frac{\pi}{2}}^{0}\sin^2\alpha\,d\alpha \Bigg] = \frac{L^2}{4}\left[\pi + \underbrace{\left(\alpha - \frac{\sin 2\alpha}{2}\right)\Big|_{\frac{\pi}{2}}^{0}}_{-\pi/2}\right] = \frac{7}{\cdot}. \text{ Therefore,}$$

$$x_c = \frac{\pi L^2}{8}/L = \frac{\pi L}{8}.$$

We can calculate the reaction force at the support, point A as $R_A = \dfrac{2\omega L}{\pi}$, and the moment $M_A = -\left(\dfrac{\pi L}{8}\right)\left(\dfrac{2\omega L}{\pi}\right) = -\dfrac{\omega L^2}{4}$, after writing the balance of forces in the z-direction and the that of the moments about point A.

<u>Shear force distribution</u>, V as a function of x is given by $V = -\int q\,dx + C_1$ where, C_1 is determined by the boundary conditions for shear force, for example $V|_{x=L} = 0$. Performing the integration gives $V = -\dfrac{2\omega}{\pi}\int \cos^{-1}\left(\dfrac{x}{L}\right)dx + C_1$. But the integral in terms of the variable α reads (see previous paragraph) $V = (-\alpha\cos\alpha + \sin\alpha) + C_1$. Applying the boundary condition at the tip of the beam, or $V|_{x=L} = V|_{\alpha=0}$ gives $C_1 = 0$. Hence, shear force as a function of x reads $V = \dfrac{2\omega L}{\pi}\left[\sqrt{1 - \dfrac{x^2}{L^2}} - \dfrac{x}{L}\cos^{-1}\left(\dfrac{x}{L}\right)\right]$. Note that the shear force equation recovers the reaction force at the support A, or $V|_{x=0} = \dfrac{2\omega L}{\pi}$.

<u>Bending moment distribution</u>, M as a function of x, is given by $M = \int V\,dx + C_2$ where, C_2 is determined by the boundary condition for moment, for example $M|_{x=L} = 0$. Performing the integration, gives $M = \dfrac{2\omega L}{\pi}\int\left(\sqrt{1 - \dfrac{x^2}{L^2}} - \dfrac{x}{L}\cos^{-1}\dfrac{x}{L}\right)dx + C_2$.

But the integral in terms of the variable α reads (recall $\cos\alpha = \dfrac{x}{L}$), $M = \dfrac{-2\omega L^2}{\pi} \int \left(\sin^2\alpha - \alpha\sin\alpha\cos\alpha \right) d\alpha + C_2$. But, using the integration by parts technique, we have $\dfrac{2\omega L^2}{\pi} \int \alpha\sin\alpha\cos\alpha\, d\alpha = \dfrac{2\omega L^2}{\pi} \left(\dfrac{\alpha\sin^2\alpha}{2} - \dfrac{1}{2}\int \sin^2\alpha\, d\alpha \right)$. Therefore, after combing the results we have

$$M = \dfrac{2\omega L^2}{\pi} \left[-\int \sin^2\alpha\, d\alpha + \dfrac{\alpha\sin^2\alpha}{2} - \dfrac{1}{2}\int \sin^2\alpha\, d\alpha + C_2 \right] = \dfrac{2\omega L^2}{\pi}$$

$\left[-\dfrac{3}{2}\int \sin^2\alpha\, d\alpha + \dfrac{\alpha\sin^2\alpha}{2} + C_2 \right]$. The remaining integral can be written as $-\dfrac{3}{2}\int \sin^2\alpha\, d\alpha = -\dfrac{3}{4}\int (1 - \cos 2\alpha)\, d\alpha = -\dfrac{3}{4}\alpha + \dfrac{\sin 2\alpha}{8}$.

Therefore, $M = \dfrac{2\omega L^2}{\pi} \left(-\dfrac{3}{4}\alpha + \dfrac{\sin 2\alpha}{8} + \dfrac{\alpha\sin^2\alpha}{2} + C_2 \right)$. Applying the boundary condition, $M\big|_{x=L} = M\big|_{\alpha=0} = 0$, we get $C_2 = 0$. Now, writing the equation for M in terms of the original variable x, we get

$$M = \dfrac{2\omega L^2}{\pi} \left(-\dfrac{3}{4}\alpha + \dfrac{\sin 2\alpha}{8} + \dfrac{\alpha\sin^2\alpha}{2} \right) = \dfrac{2\omega L^2}{\pi} \left[-\dfrac{3}{4}\cos^{-1}\left(\dfrac{x}{L}\right) + \dfrac{1}{4}\dfrac{x}{L} \right.$$

$\left. \sqrt{1 - \dfrac{x^2}{L^2}} + \dfrac{1}{2}\cos^{-1}\left(\dfrac{x}{L}\right)\left(1 - \dfrac{x^2}{L^2}\right) \right]$. After collecting similar terms in

this expression and simplifying, we have $M = \dfrac{\omega L^2}{2\pi} \left[\dfrac{x}{L}\sqrt{1 - \dfrac{x^2}{L^2}} - \right.$

$\left. \left(1 + \dfrac{2x^2}{L^2} \right)\cos^{-1}\left(\dfrac{x}{L}\right) \right]$.

14   A CANTILEVER BEAM WITH PARABOLIC LOAD DISTRIBUTION

A cantilever beam is loaded with an inverse cosine distributed load. The load is given as $q = \omega\left(1 - \dfrac{x^2}{L^2}\right)$ where, ω is the load density per unit length (e.g., N/m), and L beam length. In this section, we present the calculations for the equivalent load, its acting location x_c from the support, and the distribution of shear and bending moment along the length of this beam.

Equivalent load W, is the area under the distribution. Therefore, we can write $W = \int_0^L q\,dx = \omega\int_0^L\left(1 - \dfrac{x^2}{L^2}\right)dx$. Performing the integration gives $W = \omega\left[x - \dfrac{x^3}{3L^2}\right]_0^L = \dfrac{2\omega L}{3}$.

Centroid of the load, x_c is distance from the support where W is acting. Therefore, $x_c = \dfrac{\omega\int_0^L x\left(1 - \dfrac{x^2}{L^2}\right)dx}{\dfrac{2\omega L}{3}} = \dfrac{3\left[\dfrac{x^2}{2} - \dfrac{x^4}{4L^2}\right]_0^L}{2L} = \dfrac{3L}{8}$.

We can calculate the reaction force at the support, point A as $R_A = \dfrac{2\omega L}{3}$, and the moment $M_A = -\left(\dfrac{3L}{8}\right)\left(\dfrac{2\omega L}{3}\right) = -\dfrac{\omega L^2}{4}$, after writing the balance of forces in the z-direction and the that of the moments about point A.

Shear force distribution, V as a function of x is given by $V = -\int q\,dx + C_1$ where, C_1 is determined by the boundary conditions for shear force, for example $V|_{x=L} = 0$. Performing the integration gives $V = -\omega\int\left(1 - \dfrac{x^2}{L^2}\right)dx = -\omega\left(x - \dfrac{x^3}{3L^2}\right)$. Applying the boundary condition at the tip of the beam, or $V|_{x=L} = 0$ gives $C_1 = \dfrac{2\omega L}{3}$. Hence, the shear force as a function of x reads $V = \dfrac{2\omega L}{3} - \omega x + \dfrac{\omega}{3L^2}x^3$. Note that the shear force equation recovers the reaction force at the support A, or $V\big|_{x=0} = \dfrac{2\omega L}{3}$.

Bending moment distribution, M as a function of x, is given by $M = \int V\,dx + C_2$ where, C_2 is determined by the boundary condition for moment, for example $M\big|_{x=L} = 0$. Performing the integration, gives $M = \omega \int \left(\dfrac{2L}{3} - x + \dfrac{x^3}{3L^2} \right) dx + C_2$. But the integral reads

$\omega \int \left(\dfrac{2L}{3} - x + \dfrac{x^3}{3L^2} \right) dx = \omega \left(\dfrac{2L}{3}x - \dfrac{x^2}{2} + \dfrac{x^4}{12L^2} \right)$. Applying the boundary condition, $M\big|_{x=L} = 0$, we get $C_2 = -\dfrac{\omega L^2}{4}$. Now, writing the equation for M we get $M = \omega \left(\dfrac{1}{12L^2}x^4 - \dfrac{1}{2}x^2 + \dfrac{2L}{3}x - \dfrac{L^2}{4} \right)$.

15   A CANTILEVER BEAM WITH CIRCULAR SEGMENT CROSS-SECTION AND QUARTER-ELLIPSE LOAD DISTRIBUTION

We consider a cantilever beam with a circular segment cross-section. The beam is under a quarter elliptical distributed load. We calculate the bending and shear stresses at the point of maximum value.

In previous sections, we calculated the moment of inertia and the centroid of a circular segment. These results are repeated here for convenience, after some simplifications:

Centroid $\bar{y}_c = \dfrac{4R\sin^3\left(\dfrac{\theta}{2}\right)}{3(\theta-\sin\theta)} - R\cos\dfrac{\theta}{2}$, distance of centroid from the segment base.

Moment of inertia $I_c = \dfrac{R^4}{8}\left[\theta - \sin\theta\cos\theta - \dfrac{64\sin^6\left(\dfrac{\theta}{2}\right)}{9(\theta-\sin\theta)}\right]$, with

respect to the neutral axis at the centroid.

Also, in the previous section, we calculated the maximum bending moment and the shear force due to a quarter-ellipse load distribution. These results are repeated here for convenience:

$V_{max} = \dfrac{\pi\omega L}{4}$, $M_{max} = -\dfrac{\omega L^2}{3}$. where, ω is the maximum load density per unit length measured at the beam support location, and L beam length.

The bending stress is given as $\sigma = \dfrac{M_{max}}{I_c}y$. Where, y is the distance measured from the neutral axis in the plane of the cross section. Assuming a bending moment that puts the top fiber of the beam at the tension (i.e., negative moment), we get

$$y_{top} = R - \bar{y} = R - \frac{4R\sin^3\left(\frac{\theta}{2}\right)}{3(\theta - \sin\theta)} = \frac{4R}{3}\left[\frac{3}{4} - \frac{\sin^3\left(\frac{\theta}{2}\right)}{\theta - \sin\theta}\right]. \text{ Therefore,}$$

$$\sigma_{top} = \frac{M_{max}}{I_c}y_{top} = -\frac{4R\omega L^2}{9}\left[\frac{3}{4} - \frac{\sin^3\left(\frac{\theta}{2}\right)}{\theta - \sin\theta}\right] / \frac{R^4}{8}\left[\theta - \sin\theta\cos\theta - \right.$$

$$\left.\frac{64\sin^6\left(\frac{\theta}{2}\right)}{9(\theta - \sin\theta)}\right]. \text{ After some simplifications, we get}$$

$$\sigma_{top} = -\frac{\omega L^2}{8R^3} \cdot \frac{192(\theta - \sin\theta) - 256\sin^3\left(\frac{\theta}{2}\right)}{9(\theta - \sin\theta)(\theta - \sin\theta\cos\theta) - 64\sin^6\left(\frac{\theta}{2}\right)}. \quad \text{Similarly,}$$

the stress due to bending at the bottom fiber of the beam reads

$$\sigma_{bottom} = \frac{M_{max}}{I_c}\bar{y}_c = \frac{\omega L^2}{3}\left(\frac{4R\sin^3\left(\frac{\theta}{2}\right)}{3(\theta - \sin\theta)} - R\cos\frac{\theta}{2}\right) / \frac{R^4}{8}\left[\theta - \sin\theta\cos\theta - \right.$$

$$\left.\frac{64\sin^6\left(\frac{\theta}{2}\right)}{9(\theta - \sin\theta)}\right] = \frac{\omega L^2}{8R^3} \cdot \frac{256\sin^3\left(\frac{\theta}{2}\right) - 192(\theta - \sin\theta)\cos\frac{\theta}{2}}{9(\theta - \sin\theta)(\theta - \sin\theta\cos\theta) - 64\sin^6\left(\frac{\theta}{2}\right)}.$$

Note that the ratio of the stress at the top and the bottom of the beam cross section is exactly equal to the value of $\frac{y_{top}}{y_{bottom}}$. Or

$$\left|\frac{\sigma_{top}}{\sigma_{bottom}}\right| = \frac{y_{top}}{y_{bottom}} = \frac{3(\theta - \sin\theta) - 4\sin^3\left(\frac{\theta}{2}\right)}{4\sin^3\left(\frac{\theta}{2}\right) - 3(\theta - \sin\theta)\cos\frac{\theta}{2}}. \text{ These results con-}$$

firm that the variation of the stress is linear across the cross section of the beam.

Exercise problems

1. Calculate area, centroid, moment of inertia, polar moment of inertia for a quarter circle. Consider the axis through centroid for the calculations.

2. Calculate area, centroid, moment of inertia, polar moment of inertia for a half circle with a half circular hole. Consider the axis through centroid for the calculations.

3. Calculate area, centroid, moment of inertia, polar moment of inertia for a quarter ellipse. Consider the axis through centroid for the calculations.

4. Calculate area, centroid, moment of inertia, polar moment of inertia for a half ellipse with a half circle hole. Consider the axis through centroid for the calculations.

5. Calculate area, centroid, moment of inertia, polar moment of inertia for a half ellipse with a half ellipse hole. Consider the axis through centroid for the calculations.

6. Calculate area, centroid, moment of inertia, polar moment of inertia for a half circle with a half ellipse hole. Consider the axis through the centroid for the calculations.

7. Calculate area, centroid, moment of inertia, polar moment of inertia for a quadratic polynomial shape. Consider the axis through centroid for the calculations.

8. Calculate area, centroid, moment of inertia, polar moment of inertia for a half ellipse with a quadratic hole. Consider the axis through centroid for the calculations.

9. Calculate area, centroid, moment of inertia, polar moment of inertia for a half ellipse with a spandrel hole. Consider the axis through centroid for the calculations.

10. Calculate area, centroid, moment of inertia, polar moment of inertia for a half circle with a rectangular add on to it along the diameter. The width of the rectangle is equal to the circle radius divided by two. Consider the axis through the centroid for the calculations.

16   PROBABILITY DENSITY FUNCTIONS-PDF

Probability density functions/distributions (e.g., Normal, Weibull) are used in engineering and statistics. Usually, the integral of these functions is needed and are referred to as CDF (cumulative distribution function). In practice, engineers use standard tables for related calculations. In this section, we use the integration techniques for calculating the CDF related to Gaussian/Normal and Weibull distributions.

16.1 Normal distribution

Basic form of Normal distribution is given as $N(x) = \int e^{-x^2} dx$, also related to as Error function. Plotting the e^{-x^2} shows that it has a maximum value of 1 at $x = 0$ and symmetrically decreases about the vertical axis. This so-called bell-shape can represent many natural and industrial random data probability distributions. In this section, we show how to perform integration this function and calculate $N(x)$ for an infinite range, $x = 0, \infty$.

We can write the integral as $N \equiv \int\limits_0^\infty e^{-x^2} dx = \int\limits_0^\infty e^{-y^2} dy$. Since the answer is equivalent regardless of the variable x or y. Now, we calculate $N^2 = \int\limits_0^\infty e^{-x^2} dx \int\limits_0^\infty e^{-y^2} dy = \int\limits_0^\infty \int\limits_0^\infty e^{-(x^2+y^2)} dx dy$. This equation is valid since, for example, the value of the integral in terms of the variable y is treated as a constant for the integral in terms of x, or vice versa.

Now, we transform the integral from Cartesian (x,y) to polar coordinate (r,θ). We can write $x = r\cos\theta$ and $y = r\sin\theta$. Therefore, $x^2 + y^2 = r^2$ and $dx dy = r dr d\theta$ (i.e., the differential area elements defined in each coordinate system). This can be shown as follow:

$$dx = \frac{\partial x}{\partial r} dr + \frac{\partial x}{\partial \theta} d\theta = \cos\theta\, dr - r\sin\theta\, d\theta \text{ and } dy = \frac{\partial y}{\partial r} dr + \frac{\partial y}{\partial \theta} d\theta =$$

$\sin\theta\, dr + r\cos\theta\, d\theta$. But the Jacobian of transformation reads

$$J = \begin{vmatrix} \cos\theta & -r\sin\theta \\ \sin\theta & r\cos\theta \end{vmatrix} = r, \text{ determinant of the Jacobian matrix.}$$

Therefore, the differential elements are related through $dxdy = Jdrd\theta = rdrd\theta$.

Now we can write $\int_0^{\infty}\int_0^{\infty} e^{-(x^2+y^2)}dxdy = \int_0^{\pi/2} d\theta\int_0^{\infty} e^{-r^2} rdr$. Note that the limits of the integrals read $r = 0,\infty$ and $\theta = 0, \pi/2$, consistent with x and $y = 0,\infty$. Now we calculate this integral in terms of r, or $\int_0^{\infty} e^{-r^2} rdr$.

Let $r^2 = u \Rightarrow 2rdr = du$ and write the new integral in terms of the variable u as $\int_0^{\infty} e^{-r^2} rdr = \frac{1}{2}\int_0^{\infty} e^{-u} du = -\frac{1}{2}\left[e^{-u}\right]_0^{\infty} = -\frac{1}{2}(0-1) = \frac{1}{2}$.

Therefore, $N^2 = \int_0^{\pi/2} d\theta\int_0^{\infty} e^{-r^2} rdr = \underbrace{\frac{1}{2}\int_0^{\pi/2} d\theta}_{=1/2} = \frac{\pi}{4}$. Or $N(x) = \int_0^{\infty} e^{-x^2} dx = $

$\frac{\sqrt{\pi}}{2}$. Note that due to symmetry, $\int_{-\infty}^{\infty} e^{-x^2} dx = \sqrt{\pi}$.

The normalized form of the integral, $\hat{N}(x)$ can be calculated by dividing both side with $\frac{\sqrt{\pi}}{2}$, or $\hat{N}(x) = \frac{2}{\sqrt{\pi}}\int_0^{\infty} e^{-x^2} dx = 1$. Similarly, we have $\frac{1}{\sqrt{\pi}}\int_{-\infty}^{\infty} e^{-x^2} dx = 1$.

The standard form of the integral, $\tilde{N}(x)$ can be calculated by using the change of the variable technique. Let $x = \frac{z}{\sqrt{2}} \Rightarrow dx = \frac{dz}{\sqrt{2}}$ and the normalized form of the integral in terms of the variable z reads as $\hat{N}(z) = \frac{2}{\sqrt{2\pi}}\int_0^{\infty} e^{-\frac{z^2}{2}} dz = 1$. Similarly, we have $\frac{1}{\sqrt{2\pi}}\int_{-\infty}^{\infty} e^{-\frac{z^2}{2}} dz = 1$.

It is customary to plot the distribution about its mean μ, versus origin and scale it with its standard deviation σ, or $z = \frac{x-\mu}{\sigma}$. Mean is defined as the integral of the first moment of distribution. Or

$$\mu = \int_0^\infty xe^{-x^2}\,dx = -\left[\frac{e^{-x^2}}{2}\right]_0^\infty = \frac{1}{2}.$$ The Variance (or Standard deviation

square, σ^2) is defined as the integral of the second moment of distribution. Or $\sigma^2 = \int_0^\infty x^2 e^{-x^2}\,dx = \frac{\sqrt{\pi}}{4}$.

16.2 Weibull distribution

Weibull density function is given as $f(t) = \frac{b}{\theta}\left(\frac{t}{\theta}\right)^{b-1} e^{-\left(\frac{t}{\theta}\right)^b}$. It gives

distribution of the function f vs. time for given shape factor, b and scale factor, θ. Here we use two-parameter Weibull for our calculations. For a given set of data, the values of b and θ can be obtained to fit a given set of data. For example, for $b = 3$ Weibull and Normal distributions are closely alike.

The CFD, is $F(t) = \int_0^t f(t)\,dt$. Let $\left(\frac{t}{\theta}\right)^b = z \Rightarrow dt = \frac{\theta^b}{bt^{b-1}}\,dz$.

Therefore, writing the integral in terms of z gives

$$\int_0^t f(t)\,dt = \int_0^t \frac{b}{\theta}\left(\frac{t}{\theta}\right)^{b-1} e^{-\left(\frac{t}{\theta}\right)^b} = \int_0^t \left[\frac{b}{\theta}\left(\frac{t}{\theta}\right)^{b-1}\right]\frac{\theta^b}{bt^{b-1}}e^{-z}\,dz = \int_0^t e^{-z}\,dz.$$

Performing the integration gives $F(t) = -e^{-z} = -\left[e^{-\left(\frac{t}{\theta}\right)^b}\right]_0^t = 1 - e^{-\left(\frac{t}{\theta}\right)^b}$.

Mean and variance can be calculated as well. These quantities involve special function, Gamma [8]. The Gamma function is defined as $\Gamma(x) = \int_0^\infty t^{x-1}e^{-t}\,dt$. It can be shown that $\Gamma(x) = (x-1)\Gamma(x-1) = (x-1)!$. The mean for Weibull distribution can be worked out as $\mu = \int_{-\infty}^\infty \frac{bt}{\theta}\left(\frac{t}{\theta}\right)^{b-1} e^{-\left(\frac{t}{\theta}\right)^b} = \theta\Gamma\left(1+\frac{1}{\theta}\right)$.

Similarly, the variance $\sigma^2 = \theta^2\left[\Gamma(1+2b) - \Gamma^2(1+1/b)\right]$.

Weibull distribution has extensive applications in Reliability engineering.

REFERENCES

[1] [Online]. Available: *https://en.wikipedia.org/wiki/Lists_of_ integrals.*

[2] K. Charlwood, "Integration on Computer Algebra Systems," *The Electronic Journal of Mathematics and Technology,* vol. 2, no. 3, pp. 291-301, 2008.

[3] J. e. Sasikala, "Applications of Integral Calculus in Engineeing," *International Journal of Science, Engineering and Management (IJSEM),* vol. 2, no. 11, pp. 112-116, 2017.

[4] "University of South Carolina," [Online]. Available: *https:// people.math.sc.edu/girardi/m142/integration/100problems. pdf.* [Accessed 2022].

[5] "LibreTexts," CXone Expert knowledge management, [Online]. Available: *https://math.libretexts.org/Courses/Monroe_Community_College/MTH_211_Calculus_II/Chapter_7%3A_Techniques_of_Integration.* [Accessed 2023].

[6] "Openstax," [Online]. Available: *https://openstax.org/books/ calculus-volume-1/pages/a-table-of-integrals.* [Accessed 2023].

[7] M. Math, "Integration Bee," [Online]. Available: *https://math. mit.edu/~yyao1/integrationbee.html.* [Accessed 2023].

[8] Daniel Zwillinger (Editor), CRC Standard Mathematical Tables and Formulas, 33rd Edition, Chapman and Hall/ CRC, 2018.

[9] "engineering Fundamentals-eFunda," eFunda Inc., [Online]. Available: *efunda.com.* [Accessed 2023].

[10] D. Z. (Editor), Table of Integrals, Series, and Products, 8th Edition, Academic Press, 2014.

www.ingramcontent.com/pod-product-compliance
Lightning Source LLC
Chambersburg PA
CBHW050125240326
41458CB00122B/1428

* 9 7 8 1 6 8 3 9 2 9 6 7 3 *